Chinese Tourism in Australia

"To say that Chinese Tourism in Australia by John Connell, Phil McManus and Xuesong Ding is timely is an understatement given the role tourism has as a vital pillar of Sino-Australian relations. The history of Chinese visitation to Australia is a longstanding one, and in taking a contemporary examination of this phenomenon, authors offer new insights that link to the wide assemblage of forces acting on the return of Chinese tourists to Australia (or not), and the implications of this."

—Joseph M. Cheer, *Professor of Sustainable Tourism and Heritage, Western Sydney University, Australia*

"This exciting and easy-to-read book tells for the first time the distinctive story of Chinese tourism in Australia, as it evolved from a fixation with exotic fauna and iconic sites to embrace growing diversity, colourful sites, and the pleasures of discovering unlikely places off beaten tracks. Essential reading for tourism researchers and scholars of Chinese culture alike."

—Chris Gibson, *Professor of Human Geography, University of Wollongong, Australia*

John Connell • Phil McManus
Xuesong Ding

Chinese Tourism in Australia

Koalas, Selfies and Red Dresses

palgrave
macmillan

John Connell
School of Geosciences
University of Sydney
Camperdown, NSW 2006, Australia

Phil McManus
School of Geosciences
The University of Sydney
Camperdown, NSW 2006, Australia

Xuesong Ding
Jackson School of Geosciences
University of Texas at Austin
Austin, TX, USA

ISBN 978-981-97-2476-5 ISBN 978-981-97-2477-2 (eBook)
https://doi.org/10.1007/978-981-97-2477-2

This Palgrave Macmillan imprint is published by the registered company Springer Nature Singapore Pte Ltd.
The registered company address is: 152 Beach Road, #21-01/04 Gateway East, Singapore 189721, Singapore

Paper in this product is recyclable.

Preface and Acknowledgements

As far as we know, this is the first book on the many facets of Chinese tourism in a single destination country and which accounts for its fascinating rise, evolution and (at least COVID-19-induced) temporary fall. It provides an overview of the two decades of twenty-first century Chinese tourism in Australia, and it is primarily based on research undertaken by us in the eighteen months preceding COVID-19. In part because of COVID-19, the research and the book are rather more focused on Sydney and the nearby region than originally anticipated, however much Sydney has been the core of Chinese tourism. We have sought to keep the book as jargon free as possible and generally abjure 'big data'. We focus exclusively on visitors from the People's Republic of China, or the 'mainland', with passing references to Hong Kong (rather more distinct in the 2010s) and to Taiwan. Prices are in Australian dollars unless otherwise stated.

We are grateful for the assistance from many people along the way, including helpful feedback at conferences and seminars in the School of Geosciences at the University of Sydney. At the risk of overlooking somebody, thanks to Liping Yan and Stephen Schweinsberg for their work on research that informs Chap. 3, and to Chris Gibson for substantial contributions in relation to Chaps. 5 and 6. We thank Liping Yan for assistance with interpretation of Chinese media in Sydney; otherwise, all translations and maps were undertaken by the third author. We would also like to thank Xu Wang and Jia Guo for guidance with identifying and interpreting 'new' Chinese tourism sites.

We are grateful to Palgrave Macmillan, particularly Marion Duval, and the anonymous reviewers who have helped to improve this book. We would particularly like to thank the Bridestowe Estate and Luke David Photography for providing photographs of Bridestowe and Lake Tyrrell respectively. We are most grateful to the several contributors to Red Book who were so enthusiastic about their photographs and invocations being published. Most of all, we would like to thank the Chinese tourists who patiently answered our questions and helped us to understand that which was foreign to us. The book would have been utterly different—probably non-existent—without them.

Camperdown, NSW 2006, Australia
Austin, TX, USA
October 2023

John Connell
Phil McManus
Xuesong Ding

Contents

List of Abbreviations

ABC	Australian Broadcasting Commission
ACD	Australian Chinese Daily
ACT	Australian Capital Territory
ADS	Approved Destination Status
AI	Artificial Intelligence
ANED	Australian New Express Daily
AUKUS	The Trilateral Security Partnership Between Australia, U.K. and U.S.A.
CNTA	Chinese National Tourist Administration
DIY	Do it yourself
DT	Daily Telegraph
eWOM	Electronic word of mouth
FIT	Free and Independent Travellers
NSW	New South Wales
PRC	People's Republic of China
RV	Recreational Vehicle
SMH	Sydney Morning Herald
STD	Sing Tao Daily
VFR	Visiting Friends and Relatives
Vlog	Video blog
WA	Western Australia
WHM	Working Holiday Maker

List of Figures

List of Tables

CHAPTER 1

A New Century

At the end of the 2010s, prior to the COVID-19 pandemic and border closures, Chinese tourists represented the largest single country of origin for tourists coming to Australia, and to many other countries. Here we recount the rapid rise of Chinese tourism over no more than two decades: an era of impressive growth in numbers that seemed endless, the future seemed bright, and the structure of tourism had taken on some unique and distinctive characteristics. A new century of tourism had dawned. We thus describe and analyse tourism trajectories in these two fascinating decades as tourists emerged from group tours to begin to map out their own destinations and itineraries. While all that ground to an abrupt halt in 2020, as tourism collapsed globally, three years later in 2023 there was an upswing in Chinese visitor numbers: perhaps a prelude to a return to the dynamic Chinese international tourism that typified earlier years. Optimistically for many this may herald a return to both the 'old order' and to new directions in distinctiveness. In this book we discuss the significance of tourism in these two dynamic decades and finally speculate on how Chinese tourism may evolve in the future.

But first, in this introduction, we contextualise Chinese tourism within the wider tourism studies literature, before we assess the 'old order'. Quite conventionally, we explain the need for and value of a distinctive approach to Chinese tourism, and why and how this book fills a substantial gap in the global tourism studies literature. Our approach is designed to

J. Connell et al., *Chinese Tourism in Australia*,
https://doi.org/10.1007/978-981-97-2477-2_1

integrate geopolitical and economic changes over time with cultural understandings of Chinese communication strategies, trust and the importance of image and social standing. We explain how these factors shape the tourism experience and subsequently the tourism market and then how particular places become tourist destinations. It is necessary to move beyond soul-less large-scale exit surveys of tourists at airports to consider the geopolitical history of Chinese tourism, the emergence of the Chinese Dream (loosely President Xi's 2013 vision and invocation that young people 'should dare to dream, work assiduously to fulfil the dreams and contribute to the revitalisation of the nation'). Such dreams have received little attention in tourism studies, 'compared to other social science and area studies communities, which have grappled with such issues for decades' (Rowen, 2021, p. 897). Dreams, aspirations and visions have certainly shaped tourism.

In this book we integrate qualitative research and small-scale quantitative research to present a more holistic perspective of the origins, processes and impacts of Chinese tourism, how this has changed over time and we speculate how this may evolve in the future. In doing so we look beyond proximate and visible manifestations of Chinese tourism to consider the importance of government policy and geopolitical factors, to analyse the growing impacts of social media and influencers, and to consider how these factors may influence Chinese tourism in coming years.

A more sophisticated understanding of Chinese tourism is critical for five main reasons. First, data are limited on Chinese tourism in Australia despite the large numbers and the impact on Australian cities, the overall economy and specific tourist-oriented locations. Second, a small number of empirical studies of Chinese tourists in Australia exist but until quite recently, they have been disproportionally concentrated in regional Queensland (e.g. Zhang & Peng, 2014; Jin et al., 2016; Gardiner & Kwek, 2017) and are not representative of Australia. Third, much research has focused on outbound Chinese tourism, but as a somewhat minor destination, Australia does not figure significantly in this discussion (Leung et al., 2014; Huang & Chen, 2016). Fourth, because of an emphasis on institutional, quantitative and marketing research, the 'industry-oriented' marketing, branding, forecasting and consumer satisfaction, little research exists on qualitative aspects of Chinese tourism focused on the experiences and aspirations of Chinese tourists. Fifth, existing studies are remarkably bereft of theoretical approaches to enrich our understanding of this important phenomenon, and few take account of the ever-shifting geopolitical context.

Unless geographical and tourism research on Chinese tourism aligns with contemporary theoretical developments in tourism studies, the type of knowledge being produced is limited. Whilst there is value in extending existing knowledge, important social insights into the Chinese tourists' experience of Australia will not emerge. A deeper understanding of Chinese tourism requires approaches that privilege social practices as the entry point for theorising and conducting research, and which seek to overcome the challenges of linguistic and cultural differences, scale and governmentality in providing a qualitative research agenda for effective engagement with Chinese tourism.

Chinese tourism in Australia is particularly fascinating because of its recency, compared with other sources of visitors, and the more so because of the rapid changes in its form and context. The few early accounts of Chinese tourism in Australia, and thus immediately after the turn of the century, refer to the early days when it was mainly a guided mass tourism of package tours with tourists visiting standard iconic sites (inevitably the Sydney Harbour Bridge and Opera House) and nature ('koala-kangaroo tourism') in New South Wales (NSW), Queensland and Victoria. The book elaborates on how this quite quickly evolved into more independent tourism, sometimes conflated with individualistic tourism, with younger tourists wary of the shortfalls and constraints of group package tours, more confident with English-language issues and more knowledgeable of what they wished to see, eventually discovering and seeking out sites of particular interest to Chinese visitors. One such example is the 'Harry Potter building' at the University of Sydney, involving an intriguing combination of educational tourism, photography and performing wizardry at the site. We initially focus on the brief history of outbound Chinese tourism to Australia, the choice of Australia as a destination, followed by the emergence, evolution and efflorescence of differentiation, diversity and niches within Chinese tourism.

As numbers have rapidly grown, with Chinese tourists becoming numerically the largest group in 2017 and younger tourists—many now English-speaking—being a rapidly growing component, tourism has diversified and gone from passive to active. What is unusual and significant about the 'new' tourism, and only here now being documented, is that it is promoted almost exclusively by female influencers within Chinese social media (such as WeChat and Little Red Book/*xiaohongshu*). It provides a good example of how tourists use social media to 'search, organize, share, and annotate their travel stories and experiences through blogs and microblogs, online communities, media sharing sites and other tools in a collaborative way' (Leung et al., 2013, p. 4). That has resulted in tourists visiting sites of particular interest to Chinese visitors rather than the 'conventional' sites of the early phase. Many of these 'new sites' are places where self-photography has particular merit, chosen in part because the sites are colourful, such as the Pink Lake in Western Australia (WA), and where visitors are advised to dress appropriately to fit in ('be sure to wear a red dress here') and images are distributed in real time to communicate with others and enhance social status. This quest for uniqueness and colour has contributed to creating a distinctively new and gendered Chinese tourist geography of Australia, which we analyse in relation to conventional tourism geographies.

The rise of tourism in Australia in a sense calibrates the middle-class boom in an increasingly prosperous China, and how this new middle-class quickly became interested in international tourism and how it was permitted to become involved in a once impossible activity (see Pham et al., 2017). It reflects the emergence of international Chinese tourism, tentatively in Asia and then in Europe and elsewhere, with a 'second wave' eventually reaching Australia. Contemporary Chinese tourism has evolved from passive observation (*zou ma guan hua*: flower viewing from horseback) (*Australian New Express*, 30 April 2016)—into a range of diverse activities and individual experiences. In tracing this trajectory, we have sought to be comprehensive, in covering both English and Chinese language sources whether in journals and books, the media, and government and private sector reports, although that can never be entirely complete.

Of course, the rise in Chinese tourism to Australia also reflects circumstances in Australia, as it recognised the virtues and values of a growing tourism industry and began to engage in marketing. Australia had some advantages. Using the World Economic Forum (2019) approach to successful tourism, four pillars emerged: an Enabling Environment (including Business Environment, Safety and Security, Health and Hygiene), Tourism and Travel Policy (including International Openness, Price Competitiveness), Infrastructure (including Air Transport Infrastructure, Ground and Port, Tourist Service Infrastructure) and Natural and Cultural Resources, Australia was ranked seventh in the world in 2019 for international tourism competitiveness. Of particular importance for this book and any analysis of future tourism potential and issues, Australia ranked highly for Natural and Cultural Resources, Air Transport Infrastructure, an Enabling Environment (which included such things as utilities and infrastructure), but it ranked poorly for Price Competitiveness (World Economic Forum, 2019). In many respects this was manifest in the successes and failings of the first two decades.

The fusion of traditional Chinese Confucian values and the growing confidence of Chinese travellers based on having better English language skills, more command of technology to assist their travels and a greater sense of safety, have also influenced and given distinctiveness to the structure of Chinese tourism. Such influences contributed to the growth of independent Chinese tourism in Australia and elsewhere, and its diversification into more individual interests, especially amongst Generation Y or Millennials (roughly those people born between 1981 and 1996). Their formative years during the 1990s and 2000s included rapid advances in digital technology and social media and some immersion in the films, literature, music and fashions of a wider world. Yet, simultaneously, the same generation grew up 'in a hierarchical structure of moral and ethical obligations to the family and society', with the five Confucian virtues of benevolence, righteousness, propriety, wisdom, and trustworthiness being a foundation for life (Gardiner & Kwek, 2017, p. 498). Collectively, these values, mixed with demographic factors and melded with an evolving socialist political economy of China, explain both the growth of Chinese tourism in Australia and internationally, its changes over time: contributory factors that will surely influence the future direction of Chinese tourism.

Complex Methodologies

Much previous knowledge of Chinese tourism has come from quantitative research, including brief exit surveys at airports, expenditure reports and data collected on lengths of stay. These approaches have provided useful basic information, but they are limited in their ability to understand and analyse the actual activities of individual Chinese tourists. They are also fraught with definitional issues (and occasional cultural misunderstandings). For example, Chinese short-term visitors stay in Australia longer than other international visitors do, with the median number of nights being nine. However, this raises questions about how 'tourists' are defined, since 20% of Chinese visitors live in China but are also permanent residents of Australia, while others are business travellers (Chap. 2). Similar issues influence analysis of tourism expenditure. Consequently, as far as possible, we have sought to focus on those for whom some notion of 'holiday' is pre-eminent.

At the same time, there remains a dearth of knowledge on the social practices of the growing number of Chinese tourists; what did they actually do, what did they want to do and how did they feel about their experiences. We were aware of the increasing diversity of tourism activities, and the evolution of tour group structures, especially as we encountered tourists on a regular basis at the University of Sydney, but we were frustrated at knowing so little about their experiences. Indeed, the presence of so many Chinese tourists on our academic doorstep was a large part of the genesis of the study.

What Follows

The chapters that follow elaborate on the themes and issues we have identified above. Chapter 2 provides a brief history of the emergence and growth of Chinese tourism in Australia, positioned within the wider international context of the phenomenon of Chinese tourism. This chapter highlights the heterogeneity of Chinese tourism—in itself a phrase that suggests homogeneity—something that becomes much more evident in later chapters. Chapter 3 highlights the role of the traditional print media and marketing approaches in generating and channelling interest in, and perceptions of, Chinese tourism.

In Chap. 4 we look beyond the conventional package tours to consider newer practices of tourism. This, we argue, aligns with a new individualism and became possible as Chinese tourists became generally younger, more technologically savvy and with superior English language skills compared with the initial tourists from China. Not only are younger Chinese tourists more technologically adept, but the advances of technology that brought the Internet and smartphone, have made the newest phases of Chinese tourism in Australia quite unlike all earlier precedents. Both technologies have profoundly shaped tourism, as we show for different locations in Australia.

This is taken further in Chap. 5, where we consider the role of social media and influencers, with a focus on the specific example of the 'Harry Potter building' at the University of Sydney. Both the University and the Harry Potter Building emerged quite independent of any marketing, and beyond the management practices of the University, as spontaneous and novel niches, at the intersection of campus tourism and popular culture tourism. The emergence of this particular 'marked place' heralded a growing trend for Chinese visitors to identify other places, likewise away from iconic sites, in Sydney and beyond, of particular interest to followers.

In Chap. 6 new practices become more evident where we consider the role of images and colour in Chinese tourism and explore how these images convey messages about the individuals in the photographs and those taking the images or videos. That took place where mainly younger, middle-class female tourists created their own itineraries, sought out distinctive sites—often largely unknown to local people and other tourists—and influenced their social media followers.

The book concludes with speculations about the future extent, direction and character of Chinese tourism in Australia, as part of a renascent international phenomenon in a new age of global geopolitics, and even austerity, and suggests some contributions to tourist studies and the possibilities of future analytical directions.

Bibliography

Gardiner, S., & Kwek, A. (2017). Chinese participation in adventure tourism: A study of Generation Y International Students' perceptions. *Journal of Travel Research, 56*(4), 496–506.

Huang, S., & Chen, G. (2016). Current state of tourism research in China. *Tourism Management Perspectives, 20*, 10–18.

Jin, X., Wu, L., Becken, S., & Ding, P. (2016). How do worry, self-efficacy and coping interact? Examining Chinese tourists to Australia. *Journal of China Tourism Research, 12*(3–4), 374–393.

Leung, D., Law, R., van Hoof, H., & Buhalis, D. (2013). Social media in tourism and hospitality: A literature review. *Journal of Travel and Tourism Marketing, 30*(1–2), 3–22.

Leung, D., Li, G., Fong, L., Law, R., & Lo, A. (2014). Current state of China tourism research. *Current Issues in Tourism, 17*(8), 679–704.

Pham, T., Nghiem, S., & Dwyer, L. (2017). The determinants of Chinese visitors to Australia: A dynamic demand analysis. *Tourism Management, 63*, 268–276.

Rowen, I. (2021). Tourism studies is a geopolitical instrument: Conferences, Confucius Institutes, and 'the Chinese Dream'. *Tourism Geographies, 23*(4), 895–914.

World Economic Forum. (2019). *The travel & tourism competitiveness report 2019: Travel and tourism at a tipping point.* World Economic Forum.

Zhang, Y., & Peng, Y. (2014). Understanding travel motivations of Chinese tourists visiting Cairns. *Journal of Hospitality and Tourism Management, 21*, 44–53.

CHAPTER 2

Boom: The Rapid Growth of Chinese Tourism

This chapter charts the rapid expansion of Chinese tourism in Australia and links it to other trends in the pattern of inbound tourism to Australia. It then analyses the early days when Chinese tourism was a mass tourism of package tours and conventional destinations. It examines where the bulk of Chinese tourists came from (mainly from in and around big cities that were airline hubs), who they were and why they came. It covers an early period of mainly east coast tourism and capital cities, iconic sites (Opera House, Harbour Bridge and Great Ocean Road) and nature (koalas and kangaroos), followed by the gradual spread and diversification of tourism into new places and activities.

In this century, from a standing start, Chinese tourists in Australia steadily became the most numerous national source, outnumbering visitors from European countries, New Zealand and North America. Despite this growth, relatively little was known or recorded on who they were (and are); hence, this chapter begins the process of tracing their socio-economic status, their geographical origins and expectations of tourism, and their attitudes to and perceptions of Australia. Consequently, and somewhat remarkably, although within a few years every Australian state had devised a specific Chinese tourism strategy to tap into this market, it appears that these were based on limited knowledge of who the tourists were, who they might be and what they wanted and expected to see (see Chap. 3). While recent research has generated more nuanced understandings of Chinese

J. Connell et al., *Chinese Tourism in Australia*,
https://doi.org/10.1007/978-981-97-2477-2_2

tourists, competition between various Australian states and destinations for this expanding market has, in effect, been competing for the same market. Whether this competition itself contributes to the growth in Chinese tourism, or changes the form of this tourism, in contrast to a unified approach, of simply marketing Australia and its various attractions, is unclear.

Outbound Tourism from China

Multiple factors in China influence the extent, type and destination of outbound tourism to many global destinations, including but not limited to Australia. China is the world's largest source of outbound tourists, and in 2018 it was also the world's fourth most visited destination with 63 million foreign visitors (Anon, 2022). This domestic tourism industry is both an antecedent of international tourism and a significant component of Chinese tourism in general, with 4.1 billion domestic tourist trips in 2021, even as COVID-19 flared (Yuan & Peluso, 2023). By way of comparison, prior to the pandemic, Chinese tourists made 150 million outbound trips in 2018, an increase of 15% in a single year, although this extraordinary number was rather less than 10% of China's population (in Wei et al., 2023) and a small fraction of the domestic tourism market in China. The nation was on the move.

Mobility was transformed by the impact of COVID-19. In 2020, China's annual outbound tourist trips fell to just over 20 million, with obvious repercussions for the formerly receiving countries (Hao, 2021). This impacted many countries, lasting until January 2023, and especially other Asian countries where Chinese tourists had accounted for 30–35% of all arrivals into Thailand, Vietnam, Japan and South Korea in the summer of 2019, and 25% of visitors to the Philippines (Cash & Kongkunakornkul, 2023). Australia, where by 2019 Chinese tourists made up 15% of all international visitors (Bagshaw, 2023), was likewise affected, but less significantly since tourism was rather less important to the national economy.

When Chinese international tourism began in the mid-1980s, few destinations were visited, with the most popular destinations being nearby Hong Kong and Macau (both of which we identify separately in this book), Japan and South Korea, with Vietnam and Thailand gradually becoming relatively more popular by 2010. Thailand, Australia and France all subsequently rose in popularity relative to other destinations (Wei et al.,

2023). Culture and heritage were important factors attracting Chinese tourists to Europe, with cultural references from films creating emotional connections for outbound tourists (Croce, 2016). At all times, 'safety, short-distance, and companionship [we]re the focal points for outbound travel' (Hao, 2021).

Middle Class China

Rapid economic growth in China, especially since the 1990s, created a large and growing middle class. While the rise in tourism is conventionally linked to this 'middle class', it is actually a wealthy group, significantly above Chinese 'middle' or median households in terms of income distribution (Jing, 2017). However, most Chinese visitors are not necessarily particularly wealthy, indicating that 'Australian tourism is no longer a luxury good' (Zhang & Peng, 2014, p. 47), despite wealth, cost and special cheap airline tickets being significant influences on travel to Australia (Wu & Pearce, 2014). Nonetheless, 'when it comes to overseas travel, it has seemed that China's wealthiest people have almost no limit on how much they are willing to spend—so long as they have a good time'; thus, a package tour of about a week or ten days to Australia cost in the order of $16,000 in the mid-2010s, but that was not out of reach for many (Jing, 2017, p. 3). With the growing numbers and spending power, the Australian tourism industry was inevitably becoming increasingly interested in capturing more of the Chinese market, and that was extending beyond the basic confines of the tourism industry, with more than 40% of Chinese tourists' total expenditure in Australia being devoted to shopping (Jin et al., 2020). Australia was far from being alone in competing for and welcoming this market.

Since the 1970s China has transformed from a quite poor and somewhat ascetic country to a more evidently consumer society. Accompanying that shift, by 2012 it had become the world's largest spender on international tourism (Jin et al., 2020). To understand why China is the largest tourist generating country in the world it is necessary to look at recent changes in China's political, economic, demographic and educational development (Fu & Sutunyarak, 2023), which are also discussed in various chapters throughout this book in terms of the extent, nature and changes in tourism practices, and how they relate to the consumption of material goods, education and property.

Contemporary Chinese consumers exhibit both Western and traditional Chinese values, and shopping comprises the highest proportion of the travel expenditure of Chinese international tourists (Jin et al., 2020). Chen et al. (2018) found, however, that 'shopping' had the biggest gap between expectation and performance in their study of Chinese tour group activity in Australia, particularly around tourists being forced to shop. Shopping can be grouped into categories of clothing and accessories, health care products, cosmetics, food, drinks and souvenirs, and more utilitarian products, with fewer purchases of conventional souvenirs in large quantities being made by Chinese tourists. This can extend to other forms of tourism, such as wine tourism, that is seen as combining utilitarian purchasing, with the growth of wine consumption part of a modern Chinese lifestyle, and the values of authenticity (Gu & Huang, 2019; Jin et al., 2020). While marketing did not promote their existence, quite clearly casinos—perhaps symbols of modernity—were an attraction for some, whereas interest in Indigenous cultures has largely been absent.

Growing incomes, increased interest in and information on the wider world and a more liberal political economy were basic starting points, but time was also important. The period of holiday time available to Chinese workers was based on length of employment in a particular company. In addition to public holidays, under China's Labour Law, employees who have worked between one and 10 years are entitled to 5 days of annual leave per year, which rises to 10 days for employees who have worked for between 10 and 20 years and then 15 days for over 20 years of service, with no longer service leave provisions (Dezan Shira and Associates, 2023). Early on, travellers were more likely to travel to countries close to China, and later undertake quick organised tours to distant locations such as Australia, with a standard package tour lasting no more than ten days (Zhang & Murphy, 2009). Only older retired workers—and there were initially few of these with adequate savings—or those of independent means could travel to Australia for any length of time.

Origins and Growth

Contemporary Chinese outbound tourism began in 1984 to nearby locations, the then Special Administrative Regions of Hong Kong and Macau, with obvious Chinese influences that presented no cultural or linguistic challenges (Dai et al., 2017). It was only after 1990 that a growing number of countries acquired Approved Destination Status (ADS), with

Australia gaining that status in 1999 for tourists arriving from Beijing, Shanghai and Guangzhou (Huang & Gross, 2010). Chinese tourism in Australia was thus a wholly twenty-first century phenomenon. Not surprisingly most early Chinese tourists came from the three largest cities—Beijing, Shanghai and Guangzhou. Equally unsurprisingly, their middle class (and middle-aged) characteristics were evident in most being bureaucrats and government officials, and two thirds having a university degree (Weiler & Yu, 2006; Weiler & Chen, 2016).

Studies of early outbound Chinese tourism surprised few by revealing that, modified by incomes, 'the foreign destination's eye-catching characteristics are the main reason that attracts tourists from China'; the ability to 'offer distinguished and memorable experiences, whether in ecotourism, rural and village tourism, coastal tourism and heritage tourism' necessarily favoured some destinations (Yuan & Peluso, 2023, p. 3). More recently, the influence of movies and television on Chinese travellers' destination perception has been seen to have especially benefited tourism to Thailand and South Korea (Croce, 2016; Wen et al., 2018), while also being identified as a contributing factor to the growth of 'red tourism' within China to mark the centenary of the founding of the Chinese Communist Party (Lin, 2022). Visual perceptions and representations of destinations have become increasingly important and evident across diverse media. In 2016 Australia was one of the ten most popular tourism destinations for Chinese outbound tourists, a list that was primarily Asian and included Thailand, Japan, South Korea, USA, Indonesia, Singapore, Malaysia and Russia (China Tourism Academy, 2017), although Sydney and Melbourne were two of the top ten world cities that Chinese tourists visited outside Asia (Ipsos, 2016). In the early years, Chinese group tourists welcomed the natural scenery, distinctive animals and clean environment, whereas language challenges meant that shopping was more difficult than in later years (Wang & Davidson, 2010).

In this century Chinese tourists from the People's Republic of China (PRC) have been the most rapidly growing segment, by national origin, of the Australian tourist market and became the largest national group in 2017, passing New Zealand (Van Hoa et al., 2018). From 2013 to 2018, the number of short-term arrivals from China to Australia rose by 14.5% per year to sit at just under an annual 1.5 million (Australian Trade and Investment Commission, 2019) (Table 2.1). Numbers of visitors from Hong Kong had similarly grown and brought the total of Chinese visitors to around 1.8 million. By then, China's outbound tourism market had

Table 2.1 Australia visitor arrivals 1976–2022

Source countries	*Selected years and visitor arrivals in thousands*				
	1976	*2018*	*2019*	*2020*	*2022*
China	0.5	1,425.8	1,438.7	207.7	89.3
New Zealand	148.6	1,384.4	1,433.8	242.4	697.6
United States	75.3	782.5	818	188.7	326
UK	72.7	744	715.8	200.8	399.2
Japan	27.9	453	498.6	91.7	79.4
Singapore	9.4	436.2	478.5	58.3	295.8
India	3.6	345.5	399.3	84	303.3
Malaysia	10.3	401	384.9	44.8	85.3
Hong Kong	10.8	305.3	315.2	60.9	53.7
South Korea	1.2	303.9	280.5	53.5	72.5

Source: ABS (2018) (The total for China excludes Taiwan and Macau), ABS (2023) (The total for China excludes Taiwan and Macau) Data calculated by the authors

grown to 149 million trips, up 15% from 2017 (China Tourism Academy, 2020). However, Australia constituted just 1% of all Chinese outbound tourism, despite the rapid growth in numbers. Indeed, with the rapid growth in the overall numbers of outbound Chinese tourists in this century, Australia's share of outbound Chinese tourists actually fell from 1.2% in 2000 to 0.8% in 2016 (ABS, 2018; China Tourism Academy, 2017). Nevertheless, despite the declining proportions, rapid growth meant that past projections had typically been exceeded. Ironically, on the eve of COVID-19, projections forecast that Australian numbers would have tripled by 2026–27, now rather unlikely.

In 2018, the blunt and simplistic 'Holiday' (64%) was the dominant reason for Chinese visitors coming to NSW, and that percentage was only slightly below that in other states, followed by 'Visiting friends and relatives (VFR)' at 18% and 'education' (13%); there were identical proportions nationally (Destination NSW, 2019). During the same decade the demographic structure changed quite rapidly. In the year to June 2016 amongst Chinese visitors to NSW, 'Older non-working' (29%) was followed by 'Young or midlife single' (27%), yet, two years later, 'Young or midlife single' (30%) had become the dominant group with 'older non-working' visitors declining to 22% of the visitors (Connell & McManus, 2019).

Chinese tourist numbers increased at the same time as their demographic composition became more youthful, which was matched by a gradual shift from package tours to the rise of free-and-independent (FIT) travel, alongside an even more rapid growth in Chinese student numbers. By 2018 a quarter of all international travellers to Australia were aged between 15 and 29, and Chinese youth were the largest component (17%) of these. The transitions in numerical growth and social composition meant that there was considerable emerging scope for tourism-related businesses to target Chinese travellers and, in many contexts, a need to ensure that whatever was being marketed met Chinese cultural expectations. The rise of younger, more independent tourists, especially women, steadily changed the type of tourism experience sought, how this was sourced and how it was communicated.

Outbound Chinese Tourism to Australia

A relatively limited amount of information centres on where the Chinese tourists have come from, what is their socio-economic status, where they travel to in Australia and why they choose Australia. Even defining Chinese tourists has proved difficult, complicated by the mobility of individuals who maintain households of some kind in both Australia and China. Travel by members of 'astronaut families' (where family members are split between countries, usually with the father mainly staying in the country of origin, China, and other family members living in Australia) is further complicated by sometimes dominant interests of 'tourists' in education, real estate and commerce, rather than 'leisure' activities. Nonetheless, the proportion of education and business visitors combined declined from 34% to 23% between 2007 and 2017, while 'real' tourist numbers increased from 47% to 54% (Tourism Research Australia, 2018). It is those 'real' tourists that we focus on here.

Chinese tourists have generally been wealthier than the broad Chinese middle class (Jing, 2017), partly reflected in their disproportionately coming from the largest 'first-tier' cities (Shanghai, Beijing, Guangzhou and Shenzhen) which dominated early tourism flows (see Fig. 2.1). The establishment of visa application centres in second and third tier cities, and more travel agencies providing visa application services, meant that people no longer needed to travel to large cities to obtain visas (Chinese Tourism Academy, 2017). Following more direct flights, tourism steadily increased from these smaller cities. In 2005 direct flights were from Beijing, Shanghai

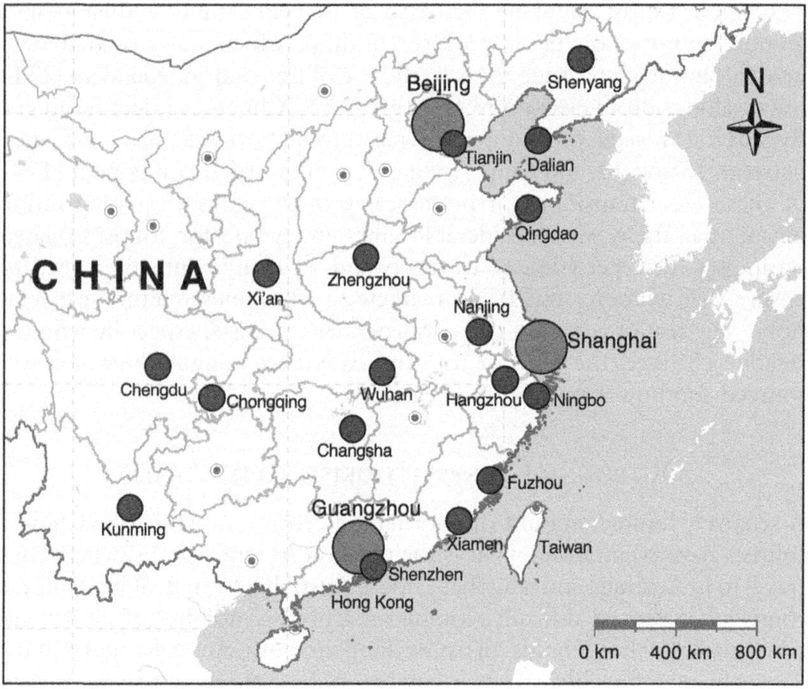

Fig. 2.1 First and second tier Chinese cities

and Guangzhou to both Sydney and Melbourne, with Chinese travellers from elsewhere using these cities, and Hong Kong, Singapore and Seoul, as transfer points to reach Australia (Zhu et al., 2019). Gradually other cities and regions were drawn in.

International flights from China also had fewer destinations in Australia. In 2010 China Southern began direct air services to Brisbane and then to Perth in 2011, so that for the first time there were direct flights beyond the east coast gateway state capitals (Zhang et al., 2021). Just as Chinese tourists came from the larger cities, so they usually flew in to the larger Australian cities (Fig. 2.2), and stayed in or near these cities as tourists. By 2016, seven Chinese airlines were providing direct flight services from ten cities in China to Australia's major capital cities, with China Southern the largest contributor to the direct connectivity between China and Australia (38%), followed by China Eastern (22%) and Air China (19%). The

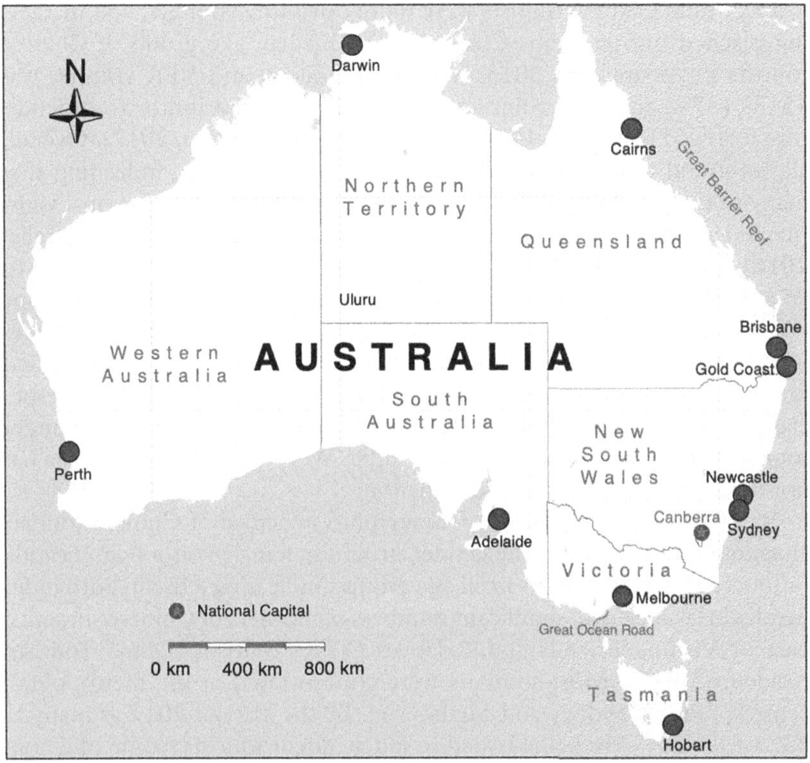

Fig. 2.2 Australian destinations

Australian-based airline Qantas contributed only 6% of the direct connectivity (Zhu et al., 2019), emphasising Chinese preference to travel on Chinese airlines. The growth in the number of flights annually between China and Australia by the big three state-owned Chinese airlines was remarkable, with Air China rising from 696 flights in 2007 to 2000 flights per year in 2017, with the corresponding figures for China Eastern being 704 flights and 3528 flights and China Southern rising from 620 flights in 2007 to 4800 flights in 2017 (Zhang et al., 2021). Newer and cheaper airlines, such as Xiamen, joined in, and the overall expansion both reflected and created demand, a peak in the industry and a dynamism abruptly ended in 2020 by COVID-19.

Over time Chinese tourists were more obviously younger, and marketing reached out to them (Chap. 3). The median age group of Chinese tourists in Australia in 2016, defined as holiday and VFR visitors, was 45–59 (37%) and this group was also the largest contributor to expenditure followed by the 30–44 age group (Tourism Australia, 2017). Overall, all outbound tourists from China are somewhat younger, indicating that many of those coming to Australia were not first-time international tourists; indeed just 3.5% were first time overseas tourists (Tourism Australia, 2016). Even by 2006 some 83% had already travelled in Asia, mostly to countries close to China, but almost half (46%) had travelled to Europe (Weiler & Yu, 2006). Thailand was familiar from television programmes while Korea and Japan offered convenient shopping, attractive dining and modern, sophisticated culture (Jing, 2017; Connell, 2018). Europe, especially France, remained the preferred overseas destination, for younger tourists with more money and leisure time (Wei et al., 2017). Australia has not therefore been a primary destination.

By 2017 not only was the demographic structure of Chinese tourists changing, but so too was the gender structure: female visitors significantly outnumbered male visitors in all age groups, indicating a rise in both independent travellers and significant numbers of mothers of Chinese students, part of Visiting Friends and Relatives (VFR) tourism (China Tourism Academy, 2017). Most students were concentrated in the metropolitan centres, notably Sydney and Melbourne. By the start of 2017 as many as 65% of all visas were being issued to independent tourists (some of whom travelled in small groups) rather than package tour members (Smith, 2016), although package tours still hold sway in many areas (Xia et al., 2018). Younger people took self-organised tours, often planned and reported via social media (Xiang & Gretzel, 2010; Wei et al., 2018). Growing numbers of competent English speakers and the availability of much web-based information supported the growth of more independent travel (Pearce et al., 2013, see Chap. 4). In the Brisbane/Gold Coast area over half of one sample were in their twenties (Jin et al., 2016) and, in areas away from bigger cities, as tourists ventured further, such as the Great Ocean Road, (Victoria) and Western Australia, they were rather younger (Han & Cheer, 2018; Xia et al., 2018).

Nonetheless, Chinese tourists over the age of 40 have exceeded those of under-40 tourists since 2012, to the extent that Bao et al. (2019) have described this as the basis of a distinctive 'third wave' (after a second wave of more sophisticated independent travellers). While this distinction may relate to different sampling methods (and different geographies), it was also the outcome of parents visiting student children, VFR tourism, older tourists choosing to visit Australia after Europe and Asia, and the continued prevalence of package tours. At the same time package tour groups were also changing in structure, with groups less likely to be characteristically of fellow workers and more likely to be of spouses and relatives (Zhang & Peng, 2014), although geographically they were still limited to the three large east coast states of NSW, Queensland and Victoria, with minimal significance in the rest of Australia (South Australian Tourism Commission, 2013). As late as 2018 the small southern island state of Tasmania, for example, received just 52,000 of the 1.3 million Chinese visitors, albeit a 37% increase on the previous year. Western Australia proportionately fared even worse, although Chinese visitors did spend more per capita than other international tourists (Ooi et al., 2021). As numbers grew and diversity increased, tourism gradually became 'a practice of relaxation-oriented vacations, fashionable accessories, and means to explore retirement locations and educational opportunities for children' (Wang et al., 2020, p. 1084). Chinese tourism had acquired multiple dimensions and more diverse destinations (Chap. 4).

In another demographic twist, as tourists were becoming younger, return visitors were becoming more frequent. By 2016 more than half (55%) of Chinese tourists in Australia were return tourists (LEK, 2017), one of the highest rates of return in Australia, comparable with return rates from Japan and USA. In 2017 only Japan had a higher rate of return. However, return was probably influenced by business interests and visits to kin. Chinese tourists may have reached Australia belatedly, but they liked what they perceived to be the natural environment, the safety, the clean air and blue sky; service provision struck a particular chord with men while women valued a natural environment more highly (Jin et al., 2020; Huang & van der Veen, 2019). Given the short time available for some travellers, many wanted to return and explore further, and many subsequently did so.

What made Chinese tourists seek to return is unclear, but may well have involved an interest in visiting more remote parts of the country. It is at least possible to conclude, as has been done, based on several surveys, that safety and shopping were important and that where memorable tourist experiences 'such as hedonism, novelty, meaningfulness, and social interaction' influenced the extent of 'fun' then intentions to recommend and revisit were stronger' (Chen et al., 2020, p. 2). Exactly what constituted these four dimensions, and whether other dimensions exist, remains unclear. Disincentives to returning are even less clear.

As the demographic composition of Chinese tourists became more youthful, and as they enthusiastically embraced modern technologies, the role of social media became ever more important in influencing the structure and process of tourism. That alone contributed to both greater individuality in tourism, and, simultaneously, the ability of tourists to follow immediate suggestions and instantaneously restructure itineraries: processes involving both short-term and long-term decisions. The role of social media in influencing Chinese tourists' travel decisions has been widely discussed (e.g. Sun et al., 2015; Wu & Pearce 2016; Du et al., 2022; Zou & Yu, 2022; Ma & Cai, 2023) but, thus far, the manner in which social media have been used to restructure routes and sites, whilst at the main destination, has been given relatively little attention, hence we consider this role later (Chap. 6). Since a single travel blog could receive hundreds of thousands of views and hundreds of reader interactions over a short period, the potential of social media to influence tourism was becoming enormous.

Diversification away from package tours was accompanied by increased per capita expenditure; Chinese tourists increasingly valued good service and spent on better accommodation and dining. Three key themes among Chinese tourists by 2016 were service experience, price-performance ratio and security (*China Youth Daily*, 28 July 2016). Chinese tourists spent $10.4 billion in Australia in 2017, which accounted for 25% of the total expenditure of all international visitors (Tourism Australia, 2018). Compared with an average expenditure of $5000 by other international tourists (LEK, 2016), two studies of Chinese tourists in Cairns and Western Australia, showed that they spent around $8000 and $4037 per visit respectively (Zhang & Peng, 2014; Xia et al., 2018), although this variation was partly a function of time, duration and methodology rather than actual expenditure differences. Such data strongly pointed, once again, to the role of shopping in Chinese tourism. While the greatest expenditure was consistently on shopping, this was generally not on purchasing conventional souvenirs, or clothes, but involved large quantities of more utilitarian products (Jin et al., 2020).

Although still focused on the east coast of Australia, mainly due to its big city attractions and airline connectivity, Chinese tourism belatedly and tentatively extended into more rural and regional areas. That was constrained by a relative lack of suitable accommodation, language, limited information and signage in Chinese, inability to use standard Chinese means of payment, including Alipay, WeChat and China Union, fewer shopping opportunities and less familiar iconic attractions (Zhang & Murphy, 2009). Since package tours were often 'short, fast-paced and spatially limited, with much time spent in Australian cities' (Weiler & Chen, 2016, p. 255), iconic sites such as the Opera House and the Harbour Bridge ensured that the 'gateway city' of Sydney was an essential part, and usually the core, of most Chinese tourist itineraries. More tourists arrived in Sydney than in other cities and few, if any, ignored these icons. However, until recently very little qualitative data has existed on Chinese tourism in Sydney (or anywhere else in Australia). Hence this book seeks to explore the growing complexities and subtleties of tourism at emerging tourist sites and contribute insights towards a better understanding of Chinese tourism in an era where the 'electronic word of mouth' (eWOM) of social media plays a significant role and various sites are singled out by the tourists themselves as 'check-in points' and 'marked places' (daka: 打卡点), which literally means 'punching the card', in effect registering a presence at 'must-visit' sites. That marked a further and greater shift from the routine of groups and packages, however much they continued to play a role, and the appearance of a more exciting, unconventional—and even unique—form.

Choosing Australia

Chinese tourism has been stimulated by similar factors to those that apply elsewhere in Asia, underpinned by the wish to see and experience a different world but one that increasingly was not entirely unfamiliar. One distinct Chinese dimension has long existed: escaping air pollution and overcrowded Chinese cities and tourist sites, with air quality and its impacts on tourism having distinctive Chinese connotations (Eusébio et al., 2021). Avoiding pollution increased the seasonality of visiting as tourists moved away from harsh, smoky winters to the Australian summer, while the natural environment and clean air also proved important influences on where and when people chose destinations (China Tourism Academy, 2017). The coastal environment (especially for Chinese inland residents), a comfortable climate, stable politics and wide-open spaces were all attractive factors, with Australia perceived as clean and 'natural', which conveyed some degree of 'authenticity'.

Travel to developed countries was regarded as a measure of quality of life, income status and well-being (China Tourism Academy, 2017; IPSOS, 2017). An early consensus suggested that the most common motivational factors for Chinese overseas travel were 'knowledge, prestige, enhancing personal relationships, relaxation, experiencing different cultures and lifestyles, and shopping' that extends into 'resting and relaxing', 'experiencing something different' and 'increasing my knowledge and experience' (Zhang & Peng, 2014, p. 45). Much depended on vantage points. One rather derisory summary suggested that Chinese tourists were 'largely not looking for either a deep cultural experience or opportunities to learn about other cultures but rather they are looking for opportunities where they can shop, play and relax' (Ruhanen et al., 2015, p. 31). However, at the same time, perceived somewhat differently, 'traditional' Chinese tourists could also be generalised as 'sightseers who hastily tick off sight by sight' (Jin & Sparks, 2017, p. 205) as they effectively marched through landscapes, museums and galleries. As one tourist reported in 2006: 'we arrived at 11pm in Melbourne on Saturday. On Sunday we did city sightseeing [usually covering about six sites] in the morning, and in the afternoon we went to Phillip Island. Today we got up very early to catch the early flight back to China' (quoted in Weiler & Yu, 2006, p. 9). There, as elsewhere, that intensity left little free time, no scope for interaction with local people and no time to participate in recreation activities.

However, in the early days of Chinese tourism, some generic notion of nature dominated, partly a function of most Australian brochures and websites then accessible in China depicting the recurrent themes of beaches, wildlife, and the Great Barrier Reef (Kwek & Lee, 2008). The growing influence of television programmes resulted in tourism to Australia being valued in terms of visible factors, natural beauty and iconic sites. Focus groups in Beijing shared perceptions of Australia that identified such generic features in more detail, especially unique animals (kangaroos and koalas) and a natural environment (fresh and clean air, sea beaches, Uluru), and specific sites (Opera House, Bondi beach, Great Barrier Reef and Harbour Bridge) concentrated on the east coast, alongside safety and a relaxed lifestyle (Huang & Gross, 2010). Indeed, despite the impressive visible symbolism of Uluru, it was not at all a key destination in the early years. Nor was there any obvious interest in the possible 'authenticity' of Australian Indigenous culture (Boag et al., 2020; Ruhanen et al., 2015). Koalas and kangaroos were pervasive in expectations even amongst VFR tourists (Austrade, 2017), but other distinctive Australian animals were almost entirely absent. Chinese online websites showed the consistent preference for beaches, architecture (iconic buildings) and distinctive animals, and thus for passive rather than active experiences (Liu et al., 2017). While such perceptions did not necessarily translate directly into tourism decisions, they indicate the broad similarity of perceptions and experiences. This early focus virtually guaranteed that the natural environment, clean air and the distinctive animals and birds were usually regarded as the most satisfactory element of the tourist experience, so that, when 'only' 20 penguins paraded at Phillip Island disappointment was marked (Weiler & Yu, 2006). Roughly a quarter of tourists have gone to see this famous Penguin Parade (Lade & Frew, 2022). Only gradually did the factors that motivated tourists begin to reflect the shift towards special interest and niche tourism, and from passive to more active tourism (Fredline & Jin, 2017; Jin & Sparks, 2017), as return visitors also became more prominent.

By the early 2000s television was becoming probably the most influential information source for Chinese travellers to Australia—emphasizing the ever-present significance of visual images. That gradually segued into the 2010s when television and film stars and influencers had a significant impact in boosting tourism in places that were hitherto less accessible to Chinese tourists, such as the Bridestowe lavender farm in Tasmania (Chap. 4). The magical Harry Potter books and movies indirectly created one specific tourism destination (Chap. 5). This highlighted the importance of factors not directly related to official tourism brochures, websites and exhibitions, but to the social worlds of ordinary Chinese citizens. Likewise, ideas and information were shared between people, either directly or increasingly through social media, rather than through the old order of static websites and travel guides. Although potential tourists remained reliant on the Internet and websites for information (74%) and rather less so on 'friends and relatives' (44%), 'social media' (38%) were steadily emerging as a source (Ipsos, 2016), as this relatively new market and younger generation became involved (Xia et al., 2018).

In the 2000s, some 30% of travellers, sampled in one survey, reported eliciting information from Chinese people in Australia (Sparks & Pan, 2009) and that percentage grew in importance. Enough Chinese residents were there to make this feasible for many, and, perhaps, it indicated both the uncertainty many early travellers had over travelling to a destination quite unlike their own, and the growth of a Chinese resident population in urban Australia (Chap. 3). Underpinning this growing interest in

Australia was its very different culture, history, and natural environment from China and the positive relationship that Australia had with China since diplomatic relationships were established early, particularly through the efforts of Gough Whitlam, the Prime Minister of Australia from 1972 to 1975 (Kim et al., 2005). Many other destinations soon established good relations with China, reducing Australia's political distinctiveness. Nevertheless, rather later, other factors included the signing of the China-Australia Free Trade Agreement (FTA) in 2015, the positive view of Australia in China relative to perceptions of other countries, and the visit of Chinese President Xi Jinping at the end of 2014. Geopolitics shaped and influenced perceptions of each other's countries, relations between them and the development of the airline infrastructure.

Chinese residents, anticipating travel to Australia, valued aesthetic experiences over entertainment, educational or escapist activities (Osmond et al., 2015), a not unusual first time, pre-trip anticipation. In the first decade of this century almost all Chinese tourists came on group package tours, which provided financial benefits, convenience and security, and focused on standard themes. This period was well summarised: 'Chinese tourists know little about Australia—at most they have some general ideas about koalas, kangaroos and the Sydney Opera House. They think of Australia as an agricultural country with flocks of sheep but otherwise a wilderness' (Pan & Laws, 2001, p. 46). Experiences largely met most such expectations and were positively regarded (Li & Carr, 2004) while tourists acquired status through having seen and photographed iconic sites and achieving basic bucket lists.

From the metropolitan centres initially limited excursions quickly took in 'standard' and accessible iconic tourist destinations such as the Blue Mountains and the Great Barrier Reef. Quite simply 'Chinese tourists have a strong desire to see the most popular and well recognized sites' (Pearce et al., 2013, p. 147). So did most other tourists. Over time tourism acquired greater diversity as word of mouth and social media became influences, and visitors increasingly opted to return. While iconic destinations have never become less than dominant, tourists ventured further to incorporate iconic sites in more distant parts of Australia, notably Uluru (formerly Ayers Rock), and the Great Ocean Road. Indicative of the shift towards social media, some 55% of independent visitors to Uluru had heard about it by personal recommendation (Austrade, 2017). That both suggested that formal marketing, in this instance, was failing, and that 'marketing' was extending beyond formal channels and taking more personal characteristics.

Australia had broad advantages for Chinese tourism, including both incidental factors such as a relatively weak dollar (versus a strong renminbi), Chinese origin residents (some being hotel, shop and tour staff) and access to Chinese food. It had strategically developed others, including ten-year multiple visas (encouraging and enabling return visits without the need to re-apply), and since 2012 more 'simple visa application', open aviation skies (resulting in many more flight connections), and more direct promotion on Chinese television. That has been enhanced by direct Chinese involvement and investment in tourism activities in Australia. Between 2014 and 2016 Chinese interests purchased eight major hotels in Melbourne and Sydney. Established businesses such as Ctrip (the largest Chinese provider of travel services including accommodation, transport, package tours and corporate travel management) and Alipay also set up branches in Australia. Chinese owned commercial ventures such as AuPake, effectively retailing a diversity of goods (Chap. 5) soon followed. These factors and increased numbers of resident Chinese, especially students, have increased familiarity with Australia, and facilitated tourism, not least of relatives of the students.

China's airlines were preferred over the airlines of other countries due to ticket prices, service and convenience; for visiting Australia, Chinese tourists thus chose airlines such as China Southern and China Eastern rather than Qantas. The introduction of an Open Skies agreement in 2016 removed capacity constraints on Chinese airlines and boosted visitor numbers. Direct aviation capacity doubled between 2011 and 2017 (LEK, 2017). Complementary seasonality offered Australia a comparative advantage, as one of the closest destinations in the southern hemisphere. Summer and Chinese New Year (from January to March) was one peak in visiting, a lesser peak being during university graduation periods (LEK, 2017). Somewhat unexpectedly, the seasonality of tourism increased over time (Ma et al., 2016).

Over time and with increased numbers, tourism has spread from a few iconic sites on the east coast. While formulaic large tour group experiences continue to dominate flows, tourism gradually involved more independent travel, increasingly of youth, to incorporate more distant, but still iconic, 'must-see' regional destinations, notably the Great Ocean Road and Uluru. That distance alone assured less passive and more active and 'upmarket' experiences. Thus, simply photographing the Harbour Bridge became supplemented by BridgeClimb (Chap. 4). Compared with an initial focus on sightseeing and shopping, Chinese tourists have shown increasing interest in culture and arts, involving some expectation of enjoying locally distinctive food, customs and experiences of local life (China Tourism Academy, 2017; World Tourism Cities Federation and Ipsos, 2016). Local seafood was especially welcome to the extent that the Sydney Fish Market was visited by more than a third of all Chinese visitors to Australia, while its popularity resulted in its being the test site for Alipay's online app-based menu translation and payment service (Ooi et al., 2021). The initial tour group experience resulting from English language limitations, safety concerns and so on, gradually extended into experimentation, so matching changes in Chinese society. The official 2019 *Annual Report on China's Outbound Tourism Development* noted that China's outbound tourists both 'Follow the crowd' (which was explained as the outcome of many outbound tourists being first time travellers and therefore group tours were still popular) and 'Follow the heart', the phrase used to describe the increasing number of outbound tourists (almost equal to tour group numbers) who travelled independently (China Tourism Academy, 2020). Although in the second decade of the twenty-first century, Chinese tourists were travelling further from the conventional east coast sites, they still usually began there. Going beyond required more time, money and a degree of independence, but diversity increased and new niches gradually evolved (see Chap. 4), arguably reflecting the presence of younger (perhaps fitter) Chinese travellers with better English language skills than older generations. The lack of English language among tourists, and of Mandarin among Australians, and the emerging technology of smartphones, may also have potentially contributed to a disproportionate emphasis on photography and visuality: scenery, performativity and colour, as discussed later (Chap. 6). Tourists can now see iconic sites, achieve more independent travel and capture and exult in their presence there. Equally significantly they began to discover and spread the word over new sites of interest (Chap. 5) enabling tourism to diverge from, and escape, Australian expectations.

New Frontiers

Although Chinese tourism globally and in Australia was dominated by organised package tours in its early years, it gradually moved more towards independent travel. Cultural and linguistic issues slowed that transition.

Initially it was structurally very similar to the late twentieth-century Japanese wave, from 1988 to 1998, when Japan provided the largest tourist numbers in Australia. The Australian industry marketed in Japan and created distinct products and signage. Japanese tourists initially came in tour groups, sought out nature (and came for weddings), centred on the east coast (where Japanese airlines arrived), and especially the Gold Coast, exhibited seasonality and came in increasing numbers because of more liberal domestic policies and the rise of a Japanese middle class (Breakey et al., 2008). They differed from the subsequent Chinese wave only in being much less involved in business and education, uninvolved in VFR and initially being less technologically sophisticated in the use of social media (impossible two decades earlier) than contemporary Chinese tourists. No doubt that perception of similarities was assisted by Australian companies failing to distinguish between the groups and assuming they would choose and undertake similar activities.

Remarkably, by the second decade of this century new Asian similarities emerged. Indian tourists, initially with smaller numbers, had become the fastest growing tourist group in Australia (Table 2.1). This third Asian wave consisted Indian tourists, represented the growing Anglophone middle class: relatively well off, involved in VFR tourism and education and centred on metropolitan destinations. Only a comparable lack of package tours, at least partly due to higher level English language skills, suggested that history was not again repeating itself.

While Chinese tourism may have been initially like earlier waves from East Asia, especially Japan, it increasingly constituted a distinct component of the contemporary Australian market, the outcome of both geopolitics and the rise of social media. At the same time, outbound Chinese tourism became an increasingly global phenomenon, influencing the environment, culture and economy of many nations, not least in Europe and Asia. In Australia these impacts are neither evenly distributed nor experienced, with larger cities having more capacity to accommodate this tourism and even minimal tourism activity potentially having a greater impact in smaller centres. Growth in numbers has accompanied a gradual social and geographical broadening of tourists' experiences, partly as numbers increased, especially of younger, independent tourists. That meant much greater technological sophistication and the use of social media which quickly had major implications for tourism (Chap. 5). Growth in numbers was recognised and accommodated in marketing campaigns that promoted Australia, or specific places and states, to potential tourists, but with only partial success, while local businesses also sought to be involved, themes we take up in the following chapter.

Bibliography

Anon. (2022). China. In *The Statesman's Yearbook 2023* (pp. 303–324). Palgrave Macmillan.

Austrade. (2017). *Visiting friends and relatives in Victoria: The China and Indian markets.* Austrade.

Australia Bureau of Statistics. (2018). *3401.0 overseas arrivals and departures, Australia.* ABS.

Australian Bureau of Statistics. (2023). *Overseas arrivals and departures, Australia: Statistics on international travel arriving in and departing from Australia.* Australian Bureau of Statistics (ABS).

Australian Trade and Investment Commission. (2019). *Australia's tourist arrivals exceed 9 million in 2018.* https://www.austrade.gov.au/news/economic-analysis/australias-tourist-arrivals-exceed-9-million-in-2018

Bagshaw, E. (2023, February 28). Tourism Australia launches campaign targeting Chinese tourists. *Sydney Morning Herald.*

Bao, J., Jin, X., & Weaver, D. (2019). Profiling the elite middle-age Chinese outbound travellers: A 3rd wave? *Current Issues in Tourism, 22*(5), 561–574.

Boag, C., Martin, R., & Bell, D. (2020). An 'authentic' Aboriginal product: State marketing and the self-fashioning of Indigenous experiences to Chinese tourists. *Australian Journal of Anthropology, 31*, 51–65.

Breakey, N., Ding, P., & Lee, T. (2008). Impact of Chinese outbound tourism to Australia. *Current Issues in Tourism, 11*(6), 587–603.

Cash, J., & Kongkunakornkul, P. (2023, April 19). Chinese tourists are back, but numbers still far from pre-COVID levels. *Reuters.* Accessed May 3, 2023, from https://www.reuters.com/world/china/chinese-tourists-are-back-numbers-still-far-pre-covid-levels-2023-04-19/

Chen, H., Weiler, B., & Young, M. (2018). Examining service shortfalls: A case study of Chinese group package tours to Australia. *Journal of Vacation Marketing, 24*(4), 371–386.

Chen, Z., Suntikul, W., & King, B. (2020). Research on tourism experiencescapes: The journey from art to science. *Current Issues in Tourism, 23*(11), 1407–1425.

China Tourism Academy. (2017). *Annual report of China outbound tourism development 2017.* Tourism Education Press. [In Chinese].

China Tourism Academy. (2020). *2019 annual report on China's outbound tourism development, Beijing.* [in Chinese]. Accessed May 3, 2023, from http://eng.ctaweb.org.cn/ctaen/c09/202103/9189e906cd044e0ebcd388693e7fb6ff.shtml

Connell, J. (2018). 'Timeless charm'? Tourism and development in Southeast Asia. In A. McGregor, L. Law, & F. Miller (Eds.), *Routledge handbook of Southeast Asian development* (pp. 153–168). Routledge.

Connell, J., & McManus, P. (2019). Flower viewing from horseback? New directions in Chinese tourism in Australia. *Australian Geographer, 50*(3), 333–347.

Croce, V. (2016). *Tourism in focus: The Chinese outbound travel market.* European Commission.

Dai, B., Jiang, Y., Yang, L., & Ma, Y. (2017). China's outbound tourism – Stages, policies and choices. *Tourism Management, 58*, 253–258.

Destination NSW. (2019). *Mainland China market profile. Year ended December 2018. For the period of January 2018 to December 2018.* Destination NSW. https://www.destinationnsw.com.au/wp-content/uploads/2019/04/mainland-china-ye-dec-18.pdf?x15361

Dezan Shira and Associates. (2023). *China briefing: Doing business in China.* https://www.china-briefing.com/doing-business-guide/china/human-resources-and-payroll/china-public-holidays-2021

Du, X., Liechty, T., Santos, C., & Park, J. (2022). 'I want to record and share my wonderful journey': Chinese Millennials' production and sharing of shortform travel videos on TikTok or Douyin. *Current Issues in Tourism, 25*(21), 3412–3424.

Eusébio, C., Carneiro, M., Madaleno, M., Robaina, M., et al. (2021). The impact of air quality on tourism: A systematic literature review. *Journal of Tourism Futures, 7*(1), 111–130.

Fredline, L., & Jin, X. (2017). The attractiveness of Australian events to Chinese visitors and barriers to attendance. *Event Management, 21*(6), 697–711.

Fu, H., & Sutunyarak, C. (2023). Demographic factors affecting Chinese tourists traveling to Thailand in the post-Covid-19 era. *Innovative Marketing, 19*(1), 162–174.

Gu, Q., & Huang, S. (2019). Profiling Chinese wine tourists by wine tourism constraints: A comparison of Chinese Australians and long-haul Chinese tourists in Australia. *International Journal of Tourism Research, 21*(2), 206–220.

Han, X., & Cheer, J. (2018). Chinese tourist mobilities and destination resilience: Regional tourism perspectives. *Asian Journal of Tourism Research, 3*(1), 159–187.

Hao, H. (2021, November 25). *Annual report on China's outbound tourism development 2021 released.* Visit Singapore. Org. https://www.visitsingapore.org/global-travel-news/annual-report-on-chinas-outbound-tourism-development-2021-released/

Huang, S., & Gross, M. (2010). Australia's destination image among Mainland Chinese travelers: An exploratory study. *Journal of Travel and Tourism Marketing, 27*, 63–81.

Huang, S., & van der Veen, R. (2019). The moderation of gender and generation in the effects of perceived destination image on tourist attitude and visit intention: A study of potential Chinese visitors to Australia. *Journal of Vacation Marketing, 25*(3), 375–389.

IPSOS. (2016). *Chinese outbound tourist consumption 2015–2016.* https://www.ipsos.com/sites/default/files/2016-06/021.1-Chinese-Outbound-Tourist.pdf

Jin, X., & Sparks, B. (2017). Barriers to offering special interest tour products to the Chinese outbound group market. *Tourism Management, 59*, 2015–2215.

Jin, X., Wu, L., Becken, S., & Ding, P. (2016). How do worry, self-efficacy and coping interact? Examining Chinese tourists to Australia. *Journal of China Tourism Research, 12*(3–4), 374–393.

Jin, H., Moscardo, G., & Murphy, L. (2020). Exploring Chinese outbound tourist shopping: A social practice framework. *Journal of Travel Research, 59*(1), 156–172.

Jing, S. (2017, January 13). For the rich, the only class is first class. *China Watch*, 3.

Kim, S., Guo, Y., & Agrusa, J. (2005). Preference and positioning analyses of overseas destinations by Mainland Chinese outbound pleasure tourists. *Journal of Travel Research, 44*, 212–220.

Kwek, A., & Lee, Y. (2008). Intra-cultural variance of Chinese tourists in destination image project: Case of Queensland, Australia. *Journal of Hospitality and Leisure Marketing, 16*(1–2), 105–135.

Lade, C., & Frew, E. (2022). The future of Chinese outbound tourism. An Australian perspective. In C. Lade & E. Frew (Eds.), *Chinese outbound tourist behaviour* (pp. 180–191). Routledge.

LEK. (2016). *Enter the Dragon: How the China tourism boom is transforming Australia.* Australia China Business Council.

LEK. (2017). *The China tourism economy: Reaching Australia's potential.* Australia China Business Council.

Li, J., & Carr, N. (2004). Visitor satisfaction: An analysis of mainland Chinese tourists on the Australian Gold Coast. *International Journal of Hospitality and Tourism Administration, 5*(3), 31–48.

Lin, X. (2022). Domestic tourism on the rise. *China Daily.* https://global.chinadaily.com.cn/a/202201/01/WS61cfbe40a310cdd39bc7ea93_1.html

Liu, Y., Bao, J., & Chen, K. (2017). Sentimental features of Chinese outbound tourists in Australia: Big-data based content analysis. *Tourism Tribune, 32*, 46–58. [In Chinese].

Ma, Y., & Cai, W. (2023). Social media in ethnographic research: Critical reflections on using WeChat in researching Chinese outbound tourists. *Current Issues in Tourism, 26*(20), 3275–3287.

Ma, E., Liu, Y., Li, J., & Chen, S. (2016). Anticipating Chinese tourists arrivals in Australia: A time series analysis. *Tourism Management Perspectives, 17*, 50–58.

Ooi, C., et al. (2021). *Realising the opportunities of Chinese tourism: A comparative study*. Asia Institute, University of Tasmania.

Osmond, A., Chen, T., & Pearce, P. (2015). Examining experience economy approaches to tourists' anticipated experiences: Mainland Chinese travellers consider Australia. *European Journal of Tourism Research, 10*, 95–108.

Pan, G., & Laws, E. (2001). Tourism marketing opportunities for Australia in China. *Journal of Vacation Marketing, 8*(1), 39–48.

Pearce, P., Wu, M., & Osmond, A. (2013). Puzzles in understanding Chinese tourist behaviour: Towards a Triple-C Gaze. *Tourism Recreation Research, 8*(2), 145–157.

Ruhanen, L., Whitford, M., & McLennan, C. (2015). Exploring Chinese visitor demand for Australia's Indigenous tourism experiences. *Journal of Hospitality and Tourism Management, 24*, 25–34.

Smith, M. (2016, 28 December). Delivering the real experience. *Australian Financial Review*, 42–43.

South Australian Tourism Commission (2013) Snapshot. (2013). *Activating China – 2020*. Government of South Australia.

Sparks, B., & Pan, G. (2009). Chinese outbound tourists: Underrating their attitudes, constraints and use of information services. *Tourism Management, 30*(4), 483–494.

Sun, M., Ryan, C., & Pan, S. (2015). Using Chinese travel blogs to examine perceived destination image: The case of New Zealand. *Journal of Travel Research, 54*(4), 543–555.

Tourism Australia. (2016). *China market profile 2015*. Tourism Australia.

Tourism Australia. (2017). *China market profile 2016*. Tourism Australia.

Tourism Australia. (2018). *International tourism snapshot as at 31 December 2017*. Tourism Australia.

Tourism Research Australia. (2018). *Trips, nights, average stay and spend by international visitors, international tourism statistics*. Australian Trade and Investment Commission. Accessed May 18, 2018.

Van Hoa, T., Turner, L., & Vu, J. (2018). Economic impact of Chinese tourism on Australia: A new approach. *Tourism Economics, 24*(6), 677–689.

Wang, Y., & Davidson, M. (2010). Pre- and post-trip perceptions: An insight into Chinese package holiday market to Australia. *Journal of Vacation Marketing, 16*(2), 111–123.

Wang, D., Kirillova, K., & Lehto, X. (2020). Tourism mobilities through time in China: A developmental and holistic lens. *Journal of Travel Research, 59*(6), 1073–1090.

Wei, X., Meng, F., & Zhang, P. (2017). Chinese Citizens' outbound destination choice: Objective and subjective factors. *International Journal of Tourism Research, 19*, 38–49.

Wei, Y., Zhang, G., & Weng, W. (2018). The influence of 'Internet Plus' on China's tourism: Evolution and breakthrough. In *Proceedings of 4th interna-*

tional conference on behavioral, economic, and socio-cultural computing (BESC) July 1, 2017, BESC (pp. 1–5).

Wei, S., Ng, S., Lee, J., & Soutar, G. (2023). Similarity-attraction cluster of outbound Chinese tourists: Who belongs there? *Journal of Hospitality & Tourism Research, 47*(1), 33–56.

Weiler, B., & Chen, H. (2016). Repositioning Australian cities as settings for memorable nature-based experiences aimed at Chinese tourists. *Tourism Recreation Research, 41*(3), 246–258.

Weiler, B., & Yu, X. (2006). *Understanding experiences of Chinese visitors to Victoria.* Sustainable Tourism Cooperative Research Centre.

Wen, H., Josiam, B. M., Spears, D. L., & Yang, Y. (2018). Influence of movies and television on Chinese Tourists perception toward international tourism destinations. *Tourism Management Perspectives, 28*, 211–219.

World Tourism Cities Federation & Ipsos. (2016). *Market research report on Chinese.*

Wu, M., & Pearce, P. (2014). Chinese recreational vehicle users in Australia: A netnographic study of tourist motivation. *Tourism Management, 43*, 22–35.

Wu, M., & Pearce, P. L. (2016). Tourism blogging motivations: Why do Chinese tourists create little 'Lonely Planets'? *Journal of Travel Research, 55*(4), 537–549.

Xia, J., Adriano, R., Zea, M., Crowe-Delaney, L., et al. (2018). *Are we China ready? Chinese tourism in Western Australia.* BCEC Research Report No. 14/18.

Xiang, Z., & Gretzel, U. (2010). Role of social media in online travel information search. *Tourism Management, 31*, 179–188.

Yuan, B., & Peluso, A. (2023, March 13). Travelling abroad or domestically? The effects of belief in karma and anxiety on post-pandemic travelling intention among Chinese tourists. *Current Issues in Tourism.*

Zhang, Y., & Murphy, P. (2009). Supply-chain considerations in marketing underdeveloped regional destinations: A case study of Chinese tourism to the Goldfields region of Victoria. *Tourism Management, 30*, 278–287.

Zhang, Y., & Peng, Y. (2014). Understanding travel motivations of Chinese tourists visiting Cairns. *Journal of Hospitality and Tourism Management, 21*, 44–53.

Zhang, Y., Zhang, C., & Yarde, K. (2021). Characteristics of China's air transport industry. In C. Zhang & K. Yarde (Eds.), *China's trade policy on international air transport: Policy goals, driving forces, and impact* (pp. 24–52). Routledge.

Zhu, Z., Zhang, A., Zhang, Y., Huang, Z., & Xu, S. (2019). Measuring air connectivity between China and Australia. *Journal of Transport Geography,* 74, 359–370.

Zou, Y., & Yu, Q. (2022). Sense of safety towards tourism destinations: A social constructivist perspective. *Journal of Destination Marketing and Management, 24*, 1007708.

CHAPTER 3

Selling Australia: Media and Marketing

While the earliest phase of Chinese tourism was tied in to 'standard' Australian and international businesses across the various components of the industry, it quite quickly became apparent to ethnic Chinese residents in Australia that they could enter and develop a valuable market sector, hence Chinese businesses oriented to the new tourism (for example through restaurants and transport businesses, and as tour guides). A domestic (Chinese) industry emerged, to some extent centred on the long-existing Chinatowns of the major metropolises, but primarily in Sydney. At the same time that enabled smaller businesses to provide more diversified possibilities enabling the first phase of 'escape' from exclusively mass tourism. As local businesses sought to participate, as tourism numbers grew, Australia and its various states and territories, were beginning to market their own attractions in a bid for greater numbers. The local print media, both Chinese and English language, watched on and broadly supported the expansion while pointing to obvious challenges. This chapter examines these trends and directions, to emphasise a rather unprepared Australia where tourist facilities (notably hotels and restaurants) were struggling to cope with a different language and culture. We examine how some of this uncertainty, alongside considerable optimism, was represented in local press coverage in both English and Chinese language media.

J. Connell et al., *Chinese Tourism in Australia*,
https://doi.org/10.1007/978-981-97-2477-2_3

Evolving Chinatowns

The early presence of existing Chinese communities in Australia, and particularly 'Chinatowns', later played a part in the development of Chinese tourism in the twenty-first Century. The first stream of Chinese migrants had arrived in Australia in the 1840s, probably 'coolies' brought in as agricultural labourers, mainly in Victoria. Their social circumstances were so dire that the Victorian government formed a select committee to examine their exploitation (Loy-Wilson, 2017). The number of Chinese migrants increased significantly in the 1850s with the gold rush that followed the discovery of gold in New South Wales and Victoria. Almost all the migrants were men, much as in gold-rush cities around the world, such as San Francisco, eventually contributing to the building of communities that were often close-knit, in part because of Anglo prejudice against the Chinese migrants.

For the rest of the century, and prior to the White Australia policy from 1901, Chinese migration was discouraged. Nonetheless by 1861 the Chinese had become the largest group of non-British migrants in Sydney, and by the turn of the century there were almost 3500 Chinese in Sydney—almost all of whom were men—who had created a now distinct Chinatown (Connell & Ip, 1981). Much the same transition occurred in Melbourne, after the demise of the gold fields, with a distinct zone of provision stores, butchers, candle-makers (and probably bakers) and opium traders (Anderson, 1990). Both Chinatowns were viewed negatively by other urban residents.

While many other Chinese lived in what was then suburban Sydney and had made a transition from mining to market gardening, more than half of those in New South Wales were in Chinatown. Businesses were emerging there with Chinese working as merchants, cabinet makers and storekeepers and by the turn of the century there were laundries and a couple of restaurants. Chinese migration was restricted further in the first half of the twentieth century and Chinese population numbers fell. By 1947 there were just 9144 Chinese in Australia (less than a third of the 1901 number). As the population fell it simultaneously became more concentrated in the various small Chinatowns, but especially that of Sydney, where Chinese cafes and restaurants began to appear in greater numbers. After 1947, with the establishment of the People's Republic of China, and the liberalisation of migration restrictions, numbers began to grow more rapidly (with more Chinese coming from other countries, notably Malaysia) and genders becoming more evenly balanced. While many migrants and residents clustered around the Chinatowns, suburbanisation again became significant, with, in Sydney, new population clusters emerging in Hurstville, Burwood and elsewhere.

Small business remained concentrated in Chinatown and by the 1970s managers, professionals and craftsmen dominated the occupations of Chinese in Sydney. Commercial development flourished there, partly because of the arrival of Chinese businessmen with capital to invest and expertise. That was accompanied by a growing acceptance of Chinese cuisine on the part of conservative Australian society coupled with the emergence of sufficient affluence to finance regular patronage of restaurants. As other economic activities declined the restaurants and, usually, associated importing businesses and stores flourished to an unprecedented degree. A number of what became prestigious restaurants were founded in that decade. On a smaller scale the same processes occurred in Melbourne.

Migration origins shifted from Hong Kong to mainland China, and increasingly focused on Sydney. By the mid-1990s, some 60% of the Chinese population of Australia lived in Sydney, and an even higher proportion were migrants from the PRC. As many new migrants moved to suburbia, and workers in the Chinese businesses commuted from elsewhere, the character of Chinatown changed, becoming much less residential, with few business owners living on their premises, although some Chinese students were moving in. Many of these students had degrees in business-oriented subjects, such as law and accountancy, and in turn they opened businesses of a different character (unlike the family organisation of existing businesses) although restaurants dominated the visible scene (Connell & Ip, 1981). Entrepreneurial ability shifted into a range of activities, some of which became significant once Chinese tourism arrived.

The significance of the growing popularity of Chinese restaurants (quite close to the central business district) brought one other significant change. Chinatown was gradually transformed from a few drab streets into a tourist destination, initially for the people of Sydney (Collins & Jordan, 2009). By the 1970s various festive activities were organised for Chinese Lunar New Year and the Moon Cake (August Moon) Festival, with the state government intervening to promote Chinatown as part of an emphasis on 'multicultural' Australia and also because it could be a useful focus for domestic tourism. Sydney's Chinatown merchants boldly appropriated the conceptual symbols of Chineseness, and ensured their part in the transformation (Anderson, 1990, p. 152). The Chinese had acquired status: Chinatown was no longer a sink of iniquity but a valuable attraction.

Further changes followed when residential tower blocks were constructed nearby, with old industrial sites cleared as part of the Darling Harbour redevelopment, typical of waterfront redevelopments elsewhere in the world, and the construction of the Sydney Entertainment Centre. Redevelopment was largely aimed at tourists, and growing numbers of Chinese students moved into the area, providing a labour force and perceiving Chinatown as 'a "natural" place in the city for food and drink and entertainment' that also provided a 'sense of home' (Wong et al., 2016, p. 9). That accelerated further as the nearby University of Technology Sydney (UTS) grew. After the mid-1970s, capital flowed into the inner city, with a boom in property investment, and over the first two decades of this century, the nearby population increased eightfold (particularly remarkable because the inner city was otherwise losing population) with Chinatown no longer defined by narrow geographical boundaries.

Meanwhile, beneath the tourism facade of ancient China, the now transformed and glitzy Chinatown, of ornamental bridges, ornate archways, lantern lights, gardens, and colourful carved dragons and temple dogs, flourished. Multiple small businesses quietly profited and became the core of businesses associated with the interrelated *daigou* trade (Chap. 4) and the expanding Chinese tourism, providing services such as transport, although retailing and restaurants remained the core. Above all restaurants offered many versions of Chinese food—from Indonesia, Malaysia, Hong Kong and all parts of China—for the new clientele: the tourists and the Chinese students (both workers and consumers). Most Chinese tourists in Sydney stayed in nearby hotels. As it was the one part of the city where businesses habitually stayed open late, it became busier than ever: almost a 'one-stop shop' for tourist wants and needs.

Chinese tourism breathed new and, to many, valuable life into Chinatown; laneways became arcades, supermarkets and food courts entered as Chinatown became commodified and institutionalised. On the fringes, in this century, Thai, Japanese and Korean restaurants emerged, so leading to the designation of a Thaitown and a Koreatown, as diaspora from various countries in Asia asserted and marketed their national identity.

While Chinatown could be understood as a fusion of Chinese culture, people and practices with western visions of Chineseness, the existence of such inner suburbs close to the centre of the largest Australian cities was invaluable for the growth of tourism. They provided restaurants and accommodation, people to work in the hospitality industry, sources of information in Chinese languages and a sense of security in environments different from what many tourists would have experienced in China or when travelling to nearby Asian destinations. In essence they were essential to the notion of being 'China Ready' that was being promulgated in various tourism offices and state capitals, as Australia (and also New Zealand) engaged in what amounted to a global competition for the huge new market, ensuring that marketing was oriented to the Chinese tourists and provided the attractions, infrastructure and services that would match their expectations (Zhang & Shelton, 2015).

No small part of the entrepreneurial success of the Chinatowns was the outcome of the continued growth of a Chinese ethnic population in the metropolitan capitals of Australia and their patronage of the restaurants and other facilities. By 2016 the census had recorded the Chinese as the fastest growing (large) migrant group in Australia (closely followed by Indians) and in 2021 the number of Australian residents of Chinese ancestry had reached 1,391,000 (some 5.5% of the population, and the third largest overseas-born group, after New Zealand the United Kingdom). Parallel growing numbers of temporary Chinese students in Australia (166,000 in early 2018) both increased familiarity with Australia and was a factor in stimulating tourism and family reunions (Austrade, 2017).

States and Marketing

While Chinatown played no formal part in marketing tourism, Australia and the various states were actively marketing Australian tourism within China, a process that began around the turn of the century as Chinese tourism began. The importance of states and marketing campaigns in promoting tourism has long been understood (Su et al., 2021). Prior to the rise of social media, it was widely assumed that tourists were following the plans and strategies of formal organisations in a top-down manner, only later being seen to involve a process of co-creation (Campos et al., 2018; Zhang et al., 2019; McManus et al., 2021), the outcome of a more subtle, sophisticated and inclusive process. As we show, in relation to the 'Harry Potter Building' at Sydney University (Chap. 5) and the emergence of *daka* destinations in Sydney and elsewhere (Chap. 6), the use of social media by tourists, and by those who anticipate becoming tourists, has contributed to inverting the top-down relationship. Formal tourism plans have not however become redundant: in fact, they proliferated rather than disappeared, and for more than a decade, they have included the use of social media as a key factor in the plan (see Tourism Western Australia, 2011). Generally the plans were oriented to increasing numbers and had few distinctive directions appropriate to Chinese visitors, perhaps symbolised in the familiarity and clichés visibly evident in the cover designs (Fig. 3.1).

As both absolute tourist numbers and per capita expenditure increased tourism marketing focused on the Chinese market, reflected also in newspaper coverage over the later years (see below). As it evolved into a much larger market, and eventually the biggest market, there was growing recognition that this was a distinct tourist market segment that consequently required an equally distinctive approach, because of language, cultural differences and expectations. That was attached to a growing realisation, as in New South Wales, that regional planning and facilities were not then up to the task of effectively accommodating the growing numbers (NSW

Fig. 3.1 State Government China tourism marketing strategies

Business Chamber, 2017). By contrast, Tasmania, without a direct airline connection to China, was struggling to receive any Chinese tourists, with 'only a few dozen visitors' who ventured that far (Feiyue, 2017). States had quite different objectives and aspirations.

From 2011 onwards every state and territory other than the Australian Capital Territory (ACT) developed specific plans for marketing themselves to the China market. Tourism Australia produced a national *China 2020 Strategic Plan* in 2011, and within a year or so Victoria, NSW, WA and Queensland had all produced plans aimed at boosting Chinese tourism. The other states and territories followed (although Tasmania alone targeted a more general Asian market). For the first time in the history of tourism marketing in Australia, states were specifically targeting visitors from a particular destination. By then marketing had taken on a distinctive cultural dimension, amplified in later years; Tourism Australia had appointed the actor, Wu Xiubo, as its ambassador in 2017—for what had been designated the China-Australia Year of Tourism, and both South Australia and Victoria have also used well-known Chinese actors and singers in promotional videos. Individual and national state efforts were supported by Austrade, the national government's trading arm, that developed and sponsored tourism trade missions to China, and by state business councils, chambers of commerce and similar groups.

Marketing also evolved both away from an east coast focus and to be targeted at more independent travellers, who were more likely to make decisions about times, places and activities. As the Managing Director of Tourism Australia pointed out in 2016: 'We are focused on targeting this independent traveller that has a high disposable income and is more inclined to seek out higher-end experiences, repeat visitation and the smaller group size because we know that is where the yields will come from' (quoted in Smith, 2016). States have also strategically, and diplomatically, sought to develop plans that did not focus only on metropolitan areas but benefited regions.

This more complex and diversified direction was universal. Tasmania had a brief boost in 2014 when President Xi visited the state but tourism quickly succumbed to the problem of indirect flights, inadequate Wi-Fi and a lack of cultural preparedness. The Northern Territory, for example, developed in 2017 a China Ready Industry Program that conventionally targeted the top of the tourism market: niche and high-end travellers was one of the cornerstones of the strategic plan, as noted by the President of the China Australia Business Council's Northern Territory branch:

> The mass market tour bus travellers—they're so 20th century. What we're looking for now is high-end travellers: big spend, low volume. They're looking for unique experiences that give them bragging rights, where they can go back home and say 'this is where I've been, you haven't been there' (ABC News, 8 February 2016).

Such an orientation made sense, especially as independent tourists spent rather more than those on group tours.

However, as Tasmania and the Northern Territory discovered, developing Chinese tourism outside the main cities proved more difficult than for other national groups. 'Australian rural destinations are not at all equipped to handle high volumes of tourists who speak only Chinese, expect to eat Chinese food and are unfamiliar with Australian cultural behaviour' (Buckley, 2014). Enthusiasm could only go so far and, indeed, at the core of the Northern Territory's plan was a poignant plea for 20 Mandarin-speaking guides.

Regional interests were also disadvantaged by the practices of the new tourism entrepreneurs. In Victoria, Chinese travel agents exhibited little interest in encouraging Chinese tourists to visit the Goldfields regions (Ballarat and Bendigo) since there were few opportunities for them to gain shopping commissions there (Zhang & Murphy, 2009) and every incentive to focus on profitable metropolitan places where commissions were possible. Nonetheless day trips did include Ballarat since Sovereign Hill, the local gold-mining historic village, had employed Mandarin speakers to explain the Chinese legacy in the gold mining operations. Day-trips to regional locations, such as the Great Ocean Road, became the usual practice to the extent that, even after more than a decade, 'barriers to remaining in the region overnight remain significant and countering the practice of tour operators containing visitor nights and expenditure in Melbourne is a formidable obstacle' (Han & Cheer, 2018, p. 171). Ironically, despite the usual significance of shopping in Chinese tourism, concern has been expressed by some Chinese tourists that tours were curtailed or reduced to a minimum to ensure time for shopping (Weiler & Chen, 2016). The need for more effective marketing and planning was evident.

Chinese tourists experienced other problems in Australia, partly related to cultural differences. Package tourists have been disappointed by culture and food, but especially with shopping. Dissatisfaction has been associated with overpricing of tourist goods, a limited range of shopping venues and opening hours, and downgrading itineraries so limiting the range of attractions. Sydney's Chinatown too has been singled out for disappointing and expensive shopping experiences (but not for the cuisine). Conversely some tourists have been disappointed with being 'forced' to shop, 'manipulation' of purchasing behaviour, limited time to see key attractions, and unhelpful and unskilled tour guides (Weiler & Yu, 2006; Wang & Davidson, 2010b; Liu et al., 2017; LEK, 2017; Chen et al., 2018). Although there is no reason to believe that Chinese tourists fared worse (or better) than others, language was almost always an additional constraint.

Despite the identification of problems, much of the growing interest from tourism development authorities was undertaken with only a limited knowledge of what Chinese tourists experience, expect and need to know, even though the need for more specific knowledge had been reiterated many times and was recognised by Chinese business interests and the media. It was at least believed that younger generations would influence the future, and that the experiences that Chinese tourists sought in Australia would change dramatically, with the millennial generation being uninterested in saving money, and wanting unique personal experiences during travel, unlike the older generation who were seemingly trapped in the tour bus, eating instant noodles, and buying in luxury Louis Vuitton stores (*Sing Tao Daily*, 18 March 2016). Nonetheless marketing has remained constrained by a lack of good data on why tourists are coming to Australia, what activities they are involved in and their broad perceptions of the tourism experience. Marketing has therefore focused primarily on business opportunities rather than tourism experiences.

While the tourism development plans suggested a primary interest in simply increasing numbers, that was underpinned by the necessity to engage in different cultural expectations. Chinese tourists were no different from tourists from other countries in wanting good hotels, free Wi-Fi, efficient service and good food but responding to cultural differences has often been singled out as a critical issue for future development. Recognising the new tourism as solutions rather than problems was not always easy. Simple distinctions include slippers in hotels and later restaurant closing hours, while the Northern Territory's Program 2016 had researched enough to recommend that hotels and motels seeking to develop Chinese tourism included such practices as a twin bedding configuration and in-room slippers, jasmine or green tea in guest rooms, bars and restaurants, China Union Pay acceptance at front desk and gift shops (where applicable), breakfasts to include congee (rice porridge), chopsticks available in hotel restaurants and menus in simplified Chinese, a Mandarin television channel, a welcome letter, and a Chinese newspaper. If hotels were able to address these problems they would qualify for $5000 support from the government (Northern Territory, 2017). Such seemingly subtle but necessary provisions, underpinned by the need for some staff to have some Chinese-speaking ability, emphasised the range of divisions that needed to be bridged, and the difficulties that faced most regional areas where Chinese speakers were few and access to Chinese television channels or newspapers difficult or impossible. Even so, metropolitan hotels were also far from having satisfactorily addressed these cultural issues.

Within the complex picture of strategies and marketing, it is apparent that various issues have been shared concerns for different states and were evident at different scales (national, state and even at the city level). By focusing on selected strategies (rather than the more limited marketing plans or the details involved in action plans), common themes and some minor differences in the approach taken by Australian governments and others to promoting Chinese tourism can be discerned (Table 3.1).

These tourism development strategies, spanning a decade, point to many similarities (and a few differences) between the states. Queensland, for example, did not focus exclusively on China but had a wider focus on eight countries plus the Hong Kong Special Administrative Region. It recognised that:

> The new Asian traveller is distinct from previous generations: they are highly affluent, educated, urbanised and connected. They prefer autonomy when travelling, rather than a structured group tour. They are also not a homogenous group—today's Asian travellers differ from country to country, city to city, and generation to generation (Tourism and Events Queensland, 2016, p. 14).

At the same time, perhaps by virtue of a longer experience of Chinese tourism than other states, Queensland's planning was rather more subtle. Tourism and Events Queensland were, for example, aware that colour was important in promotions: 'red represents happiness, good fortune and luck', gold represents wealth and 'green represents long life but a green hat can also mean your wife is having an affair!' (Tourism and Events

Table 3.1 State and territory development plans for Chinese tourism

Government	*Strategy*	*Foundation/pillars/directions*	*Challenges*
NSW (state level) released November 2012	China Tourism Strategy 2012–2020	Eight 'strategic direction', that included extending marketing activities into four primary China source markets, (Greater Beijing, Greater Shanghai, Guangdong Province and Chengdu), supporting aviation and route development, targeting high performing consumer segments (including China's emerging middle class consumers) and improving the quality and range of visitor experiences, by working with the NSW tourism industry to research customer-focused Chinese visitor needs.	These include significant changes to consumer travel preferences, geographic source markets in China, direct airline services to other states and the escalation of competitor threat.
Western Australia (state level) released 2011	Our Direction in China: 2012–2015	Five 'strategic pillars' focused on developing direct and indirect aviation services into and within WA, consumer marketing, industry development, trade development and strategic partnerships.	Fierce international competition (USA has a US$21 million budget to market to China)
Australian Capital Territory (ACT) (territory) released 2014	2020 Tourism Strategy: Growing the visitor economy 2014–2020	Not just China focused. Had 'seven key pillars' including domestic marketing innovation, focus on digital and to establish international air access and increase domestic aviation capacity (key market focus was on Singapore as a global hub to China and Europe, and on New Zealand).	The cost to get to the ACT and the cost of activities were big issues for respondents from China. Respondents from Singapore and NZ preferred to travel to other places or stated a lack of interest in visiting Canberra and the ACT.

(*continued*)

Table 3.1 (continued)

Government	*Strategy*	*Foundation/pillars/directions*	*Challenges*
Victoria (state level) released May, 2012	Victoria's China Tourism Strategy	Five priorities—strengthening Victoria's market position, building on education/migrant/business links with China, aligning tourism investment to the Chinese visitor market, enhancing Chinese visitor experiences and improving visitor access (particularly aviation)	Lack of airline seat capacity relative to tourism targets and Chinese carriers achieving business viability in the off-peak season.
Queensland (state level) Released August 2016.	Queensland Asia Tourism Strategy 2016–2025	Aspirational goal: By 2025 Queensland will be recognised as the leading Australian destination for market share, reputation and experience delivery by Asian travellers.	Competition: capital city experiences in Melbourne and Sydney and aligned nature-based experiences. Globally, Hawaii, Fiji and the Maldives with similar climate and natural environments are direct competitors for Queensland.

Sources: Destination NSW (2020), Tourism and Events Queensland (2016), Tourism Western Australia (2011), Australian Capital Territory Government (2014), State Government Victoria (2012)

Queensland, 2013, p. 6). Their plans were coloured accordingly but translating such observations into increased tourist numbers seemed unlikely.

New South Wales, having gained Chinese tourists before other states, feared competition from them (and from overseas), that would come from direct airline services to their state capitals, and also newer choices on the part of the Chinese tourists. All states were concerned over some facets of geopolitics, including trade development, and were anxious to gain direct access to air routes to China (although they had no ability to achieve this independently). Beyond that, the strategies were not particularly sophisticated: often being generic wish lists rather than actual plans, seeking to enhance visitor experiences but with little clarity on how that might be achieved. Remarkably none of the strategies made any reference to Chinatowns, perhaps on the assumption that the tourists would inevitably discover these by themselves, or perhaps the authors felt constrained to focus on 'Australian' attractions. Thus in the NSW plan, which was aimed at marketing 'Sydney's glamour, beauty, luxury and exciting major events and the easily accessible natural wonders of its surrounding destinations', Chinatown was entirely ignored and the images of Sydney in the report focused more on The Rocks (the site of first settlement on Sydney Harbour). Perhaps less remarkably, just as Chinese tourists appeared to have little interest in seeking out Indigenous Aboriginal cultures in Australia (Boag et al., 2020), so the marketing plans rarely mentioned Indigenous cultures. Nor was there any focus on involving local Chinese workers and so contributing to local employment and business development.

Beyond the efforts of Chinatown businesses to be involved in and support tourism, other parts of the private sector also focused on encouraging and meeting Chinese needs in straightforward ways. This was evident in the actions of two of Australia's largest department stores (Myers and David Jones) and a large owner of shopping centres. Flagship Myer stores have bilingual staff (identifiable by national flag icons on their badges), sell products popular with Chinese shoppers, have special promotions during Chinese New Year and accept China Union (the preferred credit card in China). Other stores enabled online payments through Alipay and WeChat. After advertising on Chinese social media sites WeChat and Weibo, David Jones created a Lunar New Year page on its website. Scentre Group (Westfield) decorated its shopping centres in red and gold, held events such as lion dancing, tea tasting and calligraphy, hired Mandarin-speaking guides and held cultural training with retailers to help them understand the needs of Chinese shoppers. The Sydney Opera House catered heavily to Chinese visitors in January and February, around Lunar New Year, with themed food and opulent floral decorations and roving performers. Such strategies were logical and, as in the Northern Territory and in the production of specific state strategies, had never hitherto been directed at a single national tourist group. For the private sector, focusing on the needs of Chinese tourists was quite new and not particularly easy.

Some specific services, such as Chemist Warehouse, were able to expand because of Chinese tourism and shopping: the Chemist Warehouse chain offered discount vouchers to Chinese visitors that were available to Alipay customers, a distinctive convenience. New economic opportunities opened up in areas where Chinese tourists congregated. In Sydney's Chinatown, former students in Australia opened distinctive restaurants and coffee-shops, while big Asian brands, such as the Beijing-origin Peking duck restaurant chain, Quanjude, opened branches (Ang & Del Bono, 2018), all at least partly oriented to tourists, and being most effective where they had Chinese staff and were linked into the *daigou* trade (Chap. 4). Australia has largely managed to avoid 'Zero-Dollar' tours, prevalent in Asia, where tour packages, at discounted prices aimed at inexperienced travellers, are characterised by poor service standards and compulsory shopping visits to merchants who pay high commissions to the operators, with most expenditure made at the source (e.g. Tse & Tse, 2015). By contrast, in Australia there were multiple local beneficiaries. Chinese stores, with different structures of local and international ownership, but with Chinese employees, whether in Chinatown or nearby, were clear beneficiaries of Chinese tourism.

The Media and Chinese Tourism

The rise of Chinese tourism inevitably attracted media interest with both Chinese language and English language media regularly commenting on developments. Here we focus on the coverage by local Sydney-based media in the late 2010s as tourism was growing particularly rapidly. Globally, 'Chinese-language newspapers and magazines, together with Chinese community associations and schools, have long been considered to be the three pillars of ethnic Chinese communities worldwide' (Gao & Zhang, 2017, p. 60), and the print media have a long history in Australia dating back to the publication of the *China Advertiser* during the Victorian Gold Rush of the 1850s (Sun, 2016). In recent decades, they have expanded in popularity with the growth in numbers of Chinese migrants to Australia. Prior to 1993, 'there were only four Chinese-language newspapers in Australia' but related to the rise in Chinese residents in Australia, 'the second half of the 2000s saw an increasing number of Chinese-language newspapers and magazines in the Chinese community' with estimates varying from approximately 30 nationwide to about 30 in Melbourne alone (Gao & Zhang, 2017, pp. 60, 72). After technological shifts they are now much less valuable sources of communication and information.

As Li et al. (2018, p. 319) have argued: 'A close analysis of media representation can locate a phenomenon in its historical, social and cultural background'. However, recognition that the Chinese-language press play an important role in the 'formation and maintenance of the political, cultural, social and economic life of the Chinese community' (Sun et al., 2011, p. 137), has rarely extended to include tourism, despite its recent rapid growth. Since a third of early travellers to Australia had previously sought information from Chinese people in Australia (Sparks & Pan, 2009), it might have been expected that the Chinese language media would have a strong interest in reporting and potentially shaping this newly emerging tourism. Likewise, since safety and risk problems have been a common travel barrier, it might have been expected that the media would have assessed any local problems.

We examined Australian newspaper coverage of Chinese tourism in Australia over 14 months, from January 2016 to February 2017, at a time of substantial growth in Chinese visitor arrivals. This period included two years of the busy Australian summer preceding the Chinese New Year peak period, which was anticipated to be a probable period for print media to report on Chinese tourism. The study covered five daily newspapers (Table 3.2) including two English language daily newspapers: *The Daily Telegraph (*DT) and *The Sydney Morning Herald* (SMH), and three Chinese language newspapers, *Australian Chinese Daily* (ACD), *Australian New Express Daily* (ANED) and *Sing Tao Daily* (STD).

Over the 14-month period, 352 news articles reporting on Chinese tourism in Australia were identified. More articles were in the Chinese language newspapers, with 103 articles in ANED, 81 articles in STD and 72 articles in ACD, compared with 64 and 32 articles in the SMH and DT respectively. Each was examined according to its story focus (the subject of the report), topic domain (whether economically, financially, socially, culturally, politically or environmentally related), content orientation (criticising, forecasting, or emphasising), and attitude or tone (positive, neutral or negative). Some themes overlapped and covered several issues.

Over 20% of the 352 articles, in each of the papers, simply reported on tourism in general with the major difference in content being the (in)frequency of reporting about hotels/restaurants/casinos, with the two Australian newspapers, the SMH (23%) and the DT (22%), consistently reporting more on this aspect of tourism than did the three Chinese language newspapers, which each averaged just 7%. The geographic coverage of the articles varied, whether in covering the destination cities or the

Table 3.2 English and Chinese daily newspapers

Title and abbreviation	*Location*	*Format*	*Frequency*	*Quarterly average printed circulation (2016)*		*Owner*	*Readership*
				Weekday	*Saturday*		
The Sydney Morning Herald (SMH)	Sydney (NSW)	Compact	Daily (Mon–Sat)	95,733	181,113	Fairfax Media	Widely read publication in Australia, mainly in Sydney and regional NSW.
Daily Telegraph (DT)	Sydney (NSW)	Tabloid	Daily (Mon–Sat)	233,837	228,360	News Limited	Very high circulation, mainly in Sydney and regional NSW.
Australian Chinese Daily (ACD)	Sydney, Metro Australia	Broadsheet	Daily (Mon–Sat)	12,626	31,142	Australian Chinese Newspapers Pty Ltd	Favoured by traditional-Chinese speaking migrants from mainland China, Taiwan, Hong Kong and Southeast Asia.
Australian New Express Daily (ANED)	Sydney, Metro Australia	Tabloid	Daily (Mon–Sat)	12,000		Kingold Media	Favoured by the Chinese speaking community who are mainly from mainland China, Singapore and Malaysia.
Sing Tao Daily (STD)	Sydney, Metro Australia	Tabloid	Daily (Mon–Sat)	15,300	25,000	Sing Tao Newspaper Group Limited	The largest Chinese-language newspaper in Australia, mainly catering to migrants from mainland China, Hong Kong and a number of Southeast Asian countries.

source regions of the tourists. In terms of Australian destination cities, the DT (47% of articles) and SMH (34%) were Sydney focused, with little attention given to destinations for Chinese tourists in other states. While the Chinese language newspapers also had a Sydney focus (ACD 26%, ANED 24% and STD 25%) they provided more geographic coverage in terms of destinations in other states, especially Queensland and Victoria (mainly Melbourne), and to a lesser extent South Australia, Western Australia and Tasmania. Where there was any interest in the sources of the tourists, it came from the Chinese press, with geographic coverage of Chinese cities usually focused on the significance of 'first tier' Chinese cities, with many reports on new flight connections between Australia and China (see Chap. 2).

Media coverage covered ten major themes (Table 3.3). As the most significant single theme, each newspaper highlighted the need to prepare for the further growth of Chinese tourism. Whilst the frequency of any theme cannot be equated with its importance in constructing a story, or the relative size and placement of that story within each publication, the absence of particular themes is telling, as are the distinctions between the newspapers. The Chinese language newspapers were more likely to discuss safety and security, attaching rather more significance to the welfare of the tourists, than did the Australian press.

Articles that focused on the growth in Chinese tourist numbers often linked the rapidly rising numbers with tourist expenditure, and opportunities to increase this expenditure using comparisons with other countries. There were straightforward suggestions that tourism was a vital industry for Australia, that China was Australia's most important source of tourists, and that any possible changes in geopolitical events or perceptions of Australia within China would damage the Australian economy. The extent to which this was overt varied considerably, with Robyn Ironside (2016, p. 21), writing in the *Daily Telegraph* under the heading 'Chinese tourism warning', that 'the Chinese government still exercises enormous control over its citizens despite the perception the country is becoming more Westernised'. Otherwise, geopolitics was largely avoided, other than in bland remarks on the value of positive Australia-China relations.

None of the newspapers had any interest in rocking the tourist boat, and preferred to avoid political issues at a time of heightened geopolitical tension. They commented on wider issues of tourism only to stress its economic significance, more obviously developing themes related to ways of profiting from the prospective future boom. In relation to airlines,

Table 3.3 Themes used in Chinese Language and English Language Australian Newspapers

Frame		*Australian Chinese Daily (%)*	*Australian New Express Daily (%)*	*Sing Tao Daily (%)*	*Sydney Morning Herald (%)*	*The Daily Telegraph (%)*
(1)	Prepare for & take advantage of growing Chinese tourism	26	29	25	27	16
(2)	Optimistic outlook	18	12	12	0	19
(3)	Understanding Chinese tourism market	10	23	4	8	0
(4)	Profit or benefit from the growing Chinese tourism	4	0	10	22	28
(5)	Importance, contribution and effect flow	4	13	1	19	9
(6)	Concerns on the development of Chinese tourism in Australia	8	4	10	11	16
(7)	Convenience and information	3	15	6	0	0
(8)	Opportunity for Chinese investors	13	9	6	2	0
(9)	Australia-China relation	3	5	4	8	6
(10)	Safety	10	3	14	2	3

growth was seen as being achieved through airline alliances and opening new routes, whereas for smaller businesses there was a focus on what companies were already doing to take advantage of, or be relevant to, this market. The theme was often about adaptation undertaken by non-Chinese companies, or businesses with little previous exposure to the Chinese tourist market.

> After advertising on Chinese social media sites WeChat and Weibo, David Jones has created a Lunar New Year landing page on its website.
> (Retailers bank on Chinese New Year spending spree, Sue Mitchell, *Sydney Morning Herald*, 8 February 2016, p. 21)

> … in order to welcome the Chinese customers, the airport implemented a series of targeted strategies, including a Mandarin-speaking airport receptionist, electronic data access and signs in simplified Chinese, acceptance of Union Pay, the Chinese lunar New Year activities, and special supplies of retail, food and drink for Chinese visitors.
> (Air China launches direct flight between Sydney and Chengdu, *Australian New Express Daily*, 27/28 Aug 2016, p. 9)

Articles were generally positive about the economic impact of Chinese tourism for Australia, Sydney and Sydney's Chinatown. The ANED sourced an article from the *Australian Financial Review* to highlight the importance of China to Australia's economy, quoting a leading financial commentator, ANZ Bank's chief economist, Warren Hogan, who estimated 'that 7% of Australia's economic growth in the last fiscal year was attributed to Chinese tourists', noting 'that tourism is Australia's largest service industry, contributing 34 billion yuan to the economy and 500,000 employees' (Li, 2016, p. 1). Stories were generally informative, bland and determinedly uncontroversial.

The *Sydney Morning Herald* focused on the Spring Holiday and Chinese New Year to link travel home within China to the growing propensity to travel by Chinese tourists to overseas destinations, including Australia. In one article, by a writer with Chinese heritage, the Chinese New Year tradition of mobility and a big family reunion was linked through domestic travel to a 'surge in overseas travel over the New Year break', which had seen 'a 57 per cent spike in visitors [to Australia] during last year's Chinese New Year period, according to Tourism Australia' and that 'Chinese visitors surpassed 1 million for the first time last year, and the country represents Australia's most valuable and fastest-growing inbound tourism market'. The rationale given included 'increasingly, those with the financial means are eschewing the annual travel rush altogether, choosing to spend the week-long holiday abroad' (Wen, 2016, p. 30). Middle class aspirations to travel, and the impact of demography, with a small nuclear family of one child now in the workforce and parents approaching retirement, were influencing travel patterns, with population change interpreted as a potential boom factor for tourism. Capturing even some of these huge domestic numbers would be invaluable.

Collectively the media articulated the perspective that understanding Chinese tourists and the tourist market would be crucial to success, but rather than consequently focus on the tourists as individuals or even groups, articles tended to focus on an anonymous and abstract tourist market. Assuredly profits would flow with the right business strategies, and possible opportunities were often suggested or hinted at:

> Tourists from Asia tend to purchase expensive watches and designer clothing, which contributes to the thriving luxury retail industry in Australia, with an estimated revenue of 1.8 billion dollars this year.
>
> (Asian visitors tend to purchase luxury watches and designer clothing, *Sing Tao Daily*, 3 August 2016, p. 11)

Such economic opportunities extended beyond the obvious confines of the tourism industry. In both the Chinese and English language newspapers, there was enthusiasm for a positive economic impact of Chinese tourism in other sectors of the economy, such as education, real estate and agriculture, offering opportunities for local businesses to promote various local products to the Chinese market. This was, however, modified by fears of competition, and recognition of competing destinations, such as Singapore. Continued and considerable visa charges were noted in both groups of newspapers.

> Why don't we cut the cost of tourist visas to a competitive figure, and make it as easy as possible for visitors to get one? That will send a strong message that Australia is serious about enlarging our visitor economy and that we are a welcoming destination.
>
> (The great visa ripoff puts up a wall between us and China, Margy Osmond, *Sydney Morning Herald*, 22 June 2016, p. 17)

> … reduce the Australia visa fee for Chinese tourists, Chinese tourists were charged 135 dollars for visa fee, while the Europeans and Americans only charged 20 dollars, meaning that the Australian treat Chinese as second-class visitors.
>
> (Chinese tourists contribute the most but we're overcharged for high visa fees, Luan Nan, *Sing Tao Daily*, 3 March 2016, p. 1)

Concern over visa fees was framed rather more as a problem for future growth in tourist numbers rather than as a problem or disincentive for the tourists themselves, but visa fees had been even more widely recognised as a distinct disincentive to tourism (and, indeed, as they were from other Asian counties) (Pham et al., 2018).

The Chinese press occasionally noted changes in contexts and systems, and reported information that might be of value for future tourists, including information on increasing numbers of flights between Australia and China, and the provision of instructions for inbound tourists. While both English language and Chinese language newspapers identified impediments to the further growth of Chinese tourism, notably the relative currency valuations, the lack of free Wi-Fi access in Australia, visa costs and the lack of signage in Chinese, the newspapers differed in how people were characterised and how perceived and potential impediments were expressed. The SMH either saw a 'crush of humanity' within China or discussed success in terms of a new generation of millennials, who spend where their parents saved, and linked this to demography (the only child inheriting the assets) and to individualism, with one example being a marketing consultant suggesting China's young millionaires (average age 38) 'don't want to be like the other 1.4 billion people in China. They want to be individuals, they want to express themselves' (Freed, 2016, p. 19). The message pointed to the need for Australian businesses to tailor services to a new, wealthier generation of Chinese millennials.

In terms of bridging cultures, the ANED was clear about quarantine issues relating to the growth in tourist numbers. The paper noted that 'some Chinese tourists violated the relevant regulations since they were unaware of the quarantine when entering Australia', with data based on procedures at Cairns International Airport where 'last year, customs authorities seized a total of 536 kilograms of smuggled/banned goods carried by Chinese tourists, many of which were [materials] for use in traditional Chinese medicine' (Xu and Wu, 2016, p. 1). The paper went on to explain the laws, the fines and the potential prison sentences that could apply. This was followed up in another issue of the same paper by Li (2016, p. 1) who wrote 'At Shanghai Pudong International Airport, 3 hours before the departure of Qantas' daily flight to Sydney, a tour guide explained the strict quarantine laws of Australia to 15 tour group members. She said: "They even have sniffer dogs. So don't lie when you fill out the form."' Several Chinese language newspapers gently focused on the need to 'do the right thing' to ensure that tourism continued to grow.

Ultimately there was some similarity between English and Chinese language newspapers in terms of themes and approaches. The Chinese language newspapers sometimes used versions of stories already run in English-language media, while they all benefited from the same media releases. The Chinese language newspapers were more likely to recognise tourist needs, the challenges experienced by them (from language and food to visa fees) and the business opportunities for local Chinese residents. They were also more likely to recognise the concerns of the tourists, from quality of service to the price of accommodation, and the cost of visas. Otherwise even stories that overtly discussed geopolitical threats and Chinese government coercion, quoting academics based in Australia, often attempted to balance the story with topics such as better understanding of Chinese tourists and the potential for business development. In short, as a number of stories noted, this highlighted the economic importance of Chinese outbound tourism to many parts of the world, and specifically in Australia. By contrast no newspaper developed a qualitative approach that might have provided vignettes of Chinese tourists and their expectations of, and experiences in, Australia. Curiously the tourists themselves were almost entirely absent; their desires, expectations and experiences remained very much hidden.

Chinese language newspapers in Australian cities have to some extent remained significant due to the continued growth in Chinese immigrants to Australia and the diversity of political perspectives and cultures resulting from such factors as the time period of migration and the regional origin (Gao & Zhang, 2017). However, that has generally not spilled over into any concerted or critical interest in Chinese tourism to Australia even in a boom period. Moreover, while the newspapers reflected topical issues, after 2016, in addition to WeChat, most Chinese-language media in Australia adopted a dual-platform delivery model, and digital Chinese-language media largely eclipsed traditional Chinese-language media in Australia in terms of reach and audience size. Indeed by 2021 both the

Sing Tao Daily and the *Australian New Express Daily* had closed down: social media had taken over.

Many stories were perhaps determinedly inoffensive, alongside a convergence of limited coverage, suggesting that apart from highlighting key issues and the economic importance of Chinese outbound tourism and the need for good relations between the two countries, the Chinese language print media in Australia did not play a key role in fostering or shaping tourism. They did, however, construct tourism as an important component of Australia's economy and were quick to remind readers of that importance. Indeed, as we show in later chapters, that role has been monopolised by social media and the interactions among Chinese tourists themselves.

Supporting Tourism Growth

Marketing is important for tourism development; attractive images and opportunities must be created. The proliferation of marketing strategies and plans focused on China in the second decade of this century were a sign that governments thought they would be effective vehicles for enhancing Chinese tourism in their jurisdictions. They were keen to be seen to be doing something in this valuable sector, highlighting the growth and importance of Chinese tourism, but also suggesting past failures and disappointments. The media consistently echoed the same themes and hinted at continued problems and disappointments. Many of the strategies were similar in that they emphasised increased aviation capacity and enhancing the visitor experience, whatever that might be, but there were some differences in strategies based on local circumstances and perceived challenges, most of which hinged on cultural differences. The plans were often vague on any realistic means of circumventing, rather than building on, these differences.

Cultural differences failed to concern the emerging entrepreneurs of Chinatown, after the hoteliers, possibly the main Australian beneficiaries of Chinese tourism, whereas even the most established retailers struggled to compete successfully, and retain the tourism trade. With, or more often without, benefits emanating from any translation of plans into practice, it was the metropolitan capital cities that benefited most, with regional interests often simply unable to engage from the sidelines (Chap. 4). With notable exceptions around visa issues, and fine words highlighting the importance of Chinese tourism and its foundation in good relations between the two countries, the Chinese and English-language media largely played a watching brief. But their role was fading fast, while, in contrast, social media was gradually playing a more dominant and instantaneous role in shaping what would become a more diverse and complex structure of Chinese tourism, themes discussed in the next two chapters.

Bibliography

ABC. (2016, February 8). Northern Territory tourism campaign targets rich Chinese, offers bragging rights, Sydney. *ABC News.*

Anderson, K. (1990). 'Chinatown Re-oriented': A critical analysis of recent redevelopment schemes in a Melbourne and Sydney enclave. *Australian Geographical Studies, 18*(2), 137–154.

Ang, I., & Del Bono, A. (2018, May 11). Sydney's Chinatown is much more of a modern bridge to Asia than a historic enclave. *The Conversation.*

Austrade. (2017). *Visiting friends and relatives in Victoria: The China and Indian markets.* Austrade.

Australian Capital Territory Government. (2014). *2020 tourism strategy: Growing the visitor economy 2014–2020.* Australian Capital Territory Government.

Boag, C., Martin, R., & Bell, D. (2020). An 'authentic' Aboriginal product: State marketing and the self-fashioning of Indigenous experiences to Chinese tourists. *Australian Journal of Anthropology, 31*, 51–65.

Buckley, R. (2014, April 25). Ecotourism for the Chinese: Up the creek without a paddle. *The Conversation.*

Campos, A., Mendes, J., Valle, P., & Scott, N. (2018). Co-creation of tourist experiences: A literature review. *Current Issues in Tourism, 21*(4), 369–400.

Chen, H., Weiler, B., & Young, M. (2018). Examining service shortfalls: A case study of Chinese group package tours to Australia. *Journal of Vacation Marketing, 24*(4), 371–386.

Collins, J., & Jordan, K. (2009). Ethnic precincts as ethnic tourism destinations in urban Australia. *Tourism Culture and Communication, 9*(1–2), 79–92.

Connell, J., & Ip, A. (1981). The Chinese in Sydney: From Chinatown to Suburbia. *Asian Profile, 9*(4), 291–308.

Destination NSW. (2020). *Sydney precinct visitor profiles.* https://www.destinationnsw.com.au/tourism/facts-andfigures/sydney-tourism-statistics/sydney-precinct-visitor-profiles

Feiyue, Y. (2017, March 23). Overture to tourists hits the high notes. *China Daily.*

Freed, J. (2016, March 18). Young Chinese prepared to spend big on travel. *Sydney Morning Herald*, 19.

Gao, J., & Zhang, L. (2017). The changing Chinese community mediascape since the early 1990s. In J. Budarick & G.-S. Han (Eds.), *Minorities and media: Producers, industries, audiences* (pp. 59–84). Macmillan.

Han, X., & Cheer, J. (2018). Chinese tourist mobilities and destination resilience: Regional tourism perspectives. *Asian Journal of Tourism Research, 3*(1), 159–187.

Ironside, R. (2016, March 18). Chinese tourism warning. *Daily Telegraph*, 21.

LEK. (2017). *The China tourism economy: Reaching Australia's potential.* Australia China Business Council.

Li, X. (2016, February 5). During the Spring Festival, a large number of Chinese tourists are expected to visit Australia. *Australian New Express Daily*, 1.

Li, J., Pearce, P. L., & Low, D. (2018). Media representation of digital-free tourism: A critical discourse analysis. *Tourism Management, 69*, 317–329.

Liu, Y., Bao, J., & Chen, K. (2017). Sentimental features of Chinese outbound tourists in Australia: Big-data based content analysis. *Tourism Tribune, 32*, 46–58. [In Chinese].

Loy-Wilson, S. (2017). Coolie alibis: Seizing gold from Chinese miners in New South Wales. *International Labor and Working Class History, 91*, 28–45.

McManus, P., Connell, J., & Ding, X. (2021). Chinese tourists at the University of Sydney: Constraints to co-creating Campus Tourism? *Current Issues in Tourism, 24*(24), 3508–3518.

Northern Territory. (2017). *China ready industry program*, mimeo.

NSW Business Chamber. (2017). *NSW visitor economy. On the move mid-term report card.* Sydney.

Pham, T., Nghiem, S., & Dwyer, L. (2018). The economic impacts of a changing visa fee for Chinese tourists to Australia. *Tourism Economics, 24*(1), 109–126.

Smith, M. (2016). Delivering the real experience. *Australian Financial Review*, 42–43.

Sparks, B., & Pan, G. (2009). Chinese outbound tourists: Underrating their attitudes, constraints and use of information services. *Tourism Management, 30*(4), 483–494.

State Government Victoria. (2012). *Victoria's China tourism strategy.* State Government Victoria.

Su, M., Wall, G., Wu, B., et al. (2021). Tourism place making through the bioluminescent 'Blue Tears' of Pingtan Islands, China. *Marine Policy, 133*, 104744.

Sun, W. (2016). *Chinese language media in Australia: Developments, challenges and opportunities.* UTS Australia-China Relations Institute.

Sun, W., Yue, A., Sinclair, J., & Gao, J. (2011). Diasporic Chinese media in Australia: A post-2008 overview. *Continuum, 25*(4), 515–527.

Tourism and Events Queensland. (2013). *Meeting the expectations of your Chinese visitors and making them feel welcome.* Tourism and Events Queensland.

Tourism and Events Queensland. (2016). *Queensland Asia tourism strategy 2016–2025.* Tourism and Events Queensland.

Tourism Western Australia. (2011). *Our direction in China 2012–2015.* Tourism Western Australia.

Tse, T., & Tse, Q. (2015). The legal aspects of 'Zero-fare' tour in shopping tourism: A case of Chinese visitors in Hong Kong. *Journal of China Tourism Research, 11*(3), 297–314.

Wang, Y., & Davidson, M. (2010b). Chinese leisure tourists: Perceptions and satisfaction with Australia. *Tourism Analysis, 14*(6), 737–747.

Weiler, B., & Chen, H. (2016). Repositioning Australian cities as settings for memorable nature-based experiences aimed at Chinese tourists. *Tourism Recreation Research, 41*(3), 246–258.

Weiler, B., & Yu, X. (2006). *Understanding experiences of Chinese visitors to Victoria*. Sustainable Tourism Cooperative Research Centre.

Wen, P. (2016, February 6). New year's revolution: Celebrating abroad. *Sydney Morning Herald*, 30.

Wong, A., Anderson, K., Ang, I., & McNeill, D. (2016). *Sydney's Chinatown in the Asian Century: From ethnic enclave to global hub*. Institute for Culture and Society, Western Sydney University.

Xu, H., & Wu, Y. (2016). Lifestyle mobility in China: Context, perspective and prospects. *Mobilities, 11*(4), 509–520.

Zhang, Y., & Murphy, P. (2009). Supply-chain considerations in marketing underdeveloped regional destinations: A case study of Chinese tourism to the Goldfields region of Victoria. *Tourism Management, 30*, 278–287.

Zhang, J., & Shelton, E. (2015). Ordering the disordered subject: A critique of Chinese outbound tourists as New Zealand seeks to become China Ready. *Tourism Analysis, 20*, 343–353.

Zhang, C., Fong, L., & Li, S. (2019). Co-creation experience and place attachment: Festival evaluation. *International Journal of Hospitality Management, 81*, 193–204.

CHAPTER 4

Follow the Heart: Smartphones and Emerging Diversity

Chinese tourism in Australia has evolved from a general group tourism almost entirely centred on key iconic sites and nature towards a more subtle and differentiated niche tourism, but that has rarely been examined in any depth. This evolution, while vital for Australia, is not confined to Australia, and has been identified by the China Tourism Academy (2020) as a distinction between 'follow the crowd' and 'follow the heart'. Touristic diversity grew alongside individual free and independent (FIT) tourism especially in the second decade of the twenty-first century, partly associated with a wave of younger tourists, more likely to be English speaking and increasingly 'tech-savvy', and thus also with the more widespread use of social media. Various niches and interests including education, real estate, surfing, wine and DIY driving emerged, although mass tourism remained substantial, notably for first time visitors. Few if any tourists devoted the entire part of their holiday to a particular niche activity, but invariably combined their preferred activities with elements of a 'standard' tour.

In this chapter we focus on the relationships between social media, individual preferences and the rise of distinct choices and preferences. In the following chapter we analyse in some detail one particular niche, educational or campus tourism, as a phenomenon of particular importance to Chinese tourism, stressing how tourism was rarely simply an end and objective in itself but had other important secondary (or even primary)

J. Connell et al., *Chinese Tourism in Australia*,
https://doi.org/10.1007/978-981-97-2477-2_4

aspects. This chapter thus examines the wider aspects of tourism and the manner in which the rise of niches brought tentative expansion away from the coast and into the regions, and to smaller places, as Chinese tourists, some attracted by marketing campaigns (Chap. 3), and others by their fellow citizens on social media, sought to 'follow the heart'.

A significant factor in the emergence of niches of various kinds was the perceived rise of individualism in Chinese society, as it opened up, and the rise in the use of social media. While social media were quickly embraced in China, more so than anywhere else in the world, there are notable differences, as some international organisations that are ubiquitous in much of the world (such as Google and YouTube) are banned in China. A distinctive alternative Chinese social media landscape has emerged, whose workings in relationship to tourism are discussed in more detail later in this book. Individual choices became more important and the social media facilitated, supported, encouraged and eventually enabled more independent travel as is elaborated on in subsequent chapters.

Enter Social Media

By 2009 China had about 338 million Internet users, a quarter of the population, with 111 million people managing a social network profile (Darragh, 2009). A year later it was claimed that there were 384 million internet users in China, with over 30 million of them 'using the internet to book tourism products, such as hotel, air ticket, and package tour which is one of the fast growth online transaction with 498 websites related to Chinese travel, the largest share of the social media was devoted to online travellers' websites, which represented 38% the social media websites. When added to the virtual communities which represented about 37% of the social media websites, the combined 'online tourists' reviews and virtual communities' websites constituted over 73% of social media (Haiyan, 2010, pp. 509, 511). This growth was not lost on Australia's tourism industry, with, for example, Tourism Western Australia (2011, p. 10) observing that 'China has become the world's biggest social media market in a relatively short time' and that not only did the then 457 million internet users there exceed the entire population of the USA, but that 'compared with the average Chinese citizen, internet users are richer, more educated and younger'. By 2015 over 90% of the rapidly increasing number of Chinese internet users were engaged in social media, primarily WeChat and SinaWeibo (Becken & Scott, 2016; see Chap. 5). These aspects of Chinese social media—its rapid take-up, the use of online reviews and virtual communities have all been significant for the evolution and structure of Chinese tourism in Australia. Crucially, more so than for any other country, they have influenced its direction in Australia.

In addition to having the world's biggest Internet user base, China also has the world's most active environment for social media: more than 700 million people use them, from blogs to social networking sites and in multiple online communities. As the world's largest internet market and digital economy, China has a vibrant and diversified social media market: 'From Taobao, Weibo, Weixin/WeChat, Didi, Toutiao to Douyin/TikTok, Kuaishou, Meituan, Bilibili, and Xiaohongshu, Chinese digital platforms have been central to people's social, cultural, and economic lives' (Yu et al., 2022, p. 254). By 2010, China's online users were spending more than 40% of their time online on social media, a figure that was continuing to grow (Chiu et al., 2012). That usage spawned an array of companies, many with technology more advanced than in the West: thus, Chinese users could embed multimedia content in social media more than 18 months before it was possible in the United States. Social media began in China in 1994 with online forums and communities and migrated to instant messaging in 1999. China's social media users not only are more active than those of other countries, but most have multiple social media accounts. The use of mobile technologies, notably smartphones (see below), to access social media is also increasingly popular in China: even in 2010 there were more than 100 million mobile social users, a number then forecast to grow by about 30% annually. Because many Chinese are somewhat sceptical and distrustful of formal institutions and authority, users disproportionately value the advice of opinion leaders and influencers in social networks, a factor that has both proved central to the evolution of Chinese tourism in Australia and has accounted for the limited role and efficacy of formal marketing.

Just as social media have become increasingly important as virtual venues for social interaction so they have influenced choice of tourism destination (Xiang & Gretzel, 2010; Nusair et al., 2013; Liu et al., 2020). A blog or a post on Facebook, Twitter, Instagram or WeChat, can receive thousands of views and reader interactions over a short time period, constituting electronic word-of-mouth (eWOM) networks that prove unusually influential in restructuring the geography of tourism. Tourists effectively undertake their own organic marketing, as 'online brand advocates' and 'unpaid influencers' (Abidin, 2016; Wilk, 2019). Among the many forms of tourism propelled by social media is visitation to sites of popular culture such as music, film and television (see Chap. 5). Over time that evolved into the creation and discovery of distinctive sites (*daka*) deemed particularly important for Chinese visitors, outside the usual ambit of tourists and invisible in standard guidebooks.

Social media have changed the way that tourists search for information, or incidentally access information, about destinations, routes, accommodation and other services (Litvin et al., 2008; Chu et al., 2020; Pourfakhimi et al., 2020). That has been particularly so for Chinese tourists, frequently dubious of formal authority, and having little use for guidebooks (and other sources) written in English. Individuals recorded their own stories, reflections and recommendations, often in situ. Social media, and thus the tourists themselves, played an increasingly integral role in creating elements of special interest tourism (or niche tourism), as spin-offs from the greater orthodoxies and past certainties of guidebooks. Influencers' posts—and the hundreds of posts by social media users without fanfare—have shaped tourists' decisions on where to go and what to look at, involving the rapid creation and interactive exchange of content.

EWOM, that part of social media loosely defined as 'all informal communications directed at consumers through Internet-based technology related to the usage or characteristics of particular goods and services' (Litvin et al., 2008, p. 461), is distinctive for tourism in being not merely contemporary but effectively instantaneous, the immediate response to personal travel experiences. Here the 'goods and services' refer to particular places, and the communications are between tourists and potential tourists. This is especially valuable where those who are posting information are similar to those who are reading it, constituting a virtual community and network, where the information is more likely to be regarded as reliable, unbiased, trustworthy and relevant.

Research on the role and impact of eWOM has hitherto largely focused on the marketing of consumer goods and hotels rather than places, and user response to eWOM from hotels and other service providers (Sotiriadis & van Zyl, 2013; King et al., 2014; Zeng & Gerritsen, 2014; McCartney & Pek, 2018; Zarezadeh et al., 2018; Sann et al., 2021). That constitutes a broadly commercial and functional focus, analogous to the straightforward role of Trip Advisor (Prayag et al., 2018). We ignore here the psychological or commercial impact of eWOM, but seek to examine ethnographically the iterative relationships unfurling between social media practice, sharing of tourist experiences online during and after a trip, and the emergence of new touristic landscapes.

Online social travel networks have become a popular platform for users to create, share and exchange information, plan itineraries, evaluate destinations, and provide recommendations (Chung & Buhalis, 2008; Zeng & Gerritsen, 2014; Xiang & Gretzel, 2010). While social media have been used to promote destinations, that function has traditionally been undertaken by national and regional authorities, rather than by individual tourists, hence may centre on entire countries rather than specific sites (Sun et al., 2015). Tourist 'consumers' value more the knowledge and experience of others involved in similar activities rather than commercial marketing messages. Old binaries of 'production' and 'consumption' dissolve in contexts where places are generated as new destinations by visitors operating outside formal industry circuits. Consequently, we initially examine the role of social media at the scale of a single site within a metropolitan area, and through the central involvement of the tourists themselves, sometimes given the ugly label 'prosumers', who both consume and produce (Ettlinger, 2019), and whose vernacular social media practices generate new urban tourism geographies.

Confounding existing tourism theories is the degree to which social media practices suffusing tourism experience are culturally inflected and intermediated. Here, distinctive social media practices among Chinese tourists mark places as interesting and photogenic destinations in a manner unknown to others, reducing information challenges whilst travelling (Jin et al., 2016). At the end of 2016, the number of registered internet users in China totalled 770 million. More significantly, mobile phone usage had reached 1.3 billion (roughly the same as the population) and more than half of these were using the internet. More than 700 million people in China thus actively engaged with mobile internet, blogs, social networking sites, and other online communities (Keane, 2017). That usage has subsequently increased and Chinese tourists have an exceptionally high level of social media use. Meanwhile the blocking of many popular Western social media sites contributed to the rapid rise of distinctly Chinese social networking sites such as WeChat (*Weixin*), SinaWeibo and RenRen.

WeChat, a Chinese multipurpose app launched in 2011, is one of China's and the world's most popular social networking programmes, becoming the world's biggest standalone app in 2018, with over 1 billion monthly active users. It allows people to send messages by text, voice and video as well as make payments, transfer money, play games, order taxis, and book tickets for hotels and travel (Harwit, 2017). Studies of its use are limited, notably in Australia, despite its recognised value for obtaining rich qualitative data (Ma et al., 2020; Skavronskaya et al., 2019). WeChat is a multi-purpose messaging, social media and mobile payment app originally released in 2011 and described as one of the most powerful apps by Forbes; it has been argued to 'speak to Chinese collectivistic cultural aspects' (Ma et al., 2020). Similar apps include Weibo, a Chinese microblogging website launched in 2009 that became a popular social media platform, *Mafengwo* and *qiongyou* (both of which are more like TripAdvisor, with a focus on travel), and bilibili and vlog which have become popular as platforms for sharing videos. *Mafengwo*, for example, started in 2006 as a travel forum for independent travellers, is considered the 'travel bible' by younger Chinese netizens, and is the primary social-media-travel platform in China with a user base of over 100 million people. *Mafengwo* is the largest platform for user-generated content in China, with over a million users and around 3 million visits per day (Wei & Wu, 2021). However, WeChat remains the most popular communications platform in the world for Chinese speakers, an 'app for everything' used by a billion people and extremely popular with Chinese residents overseas. Thus in the United States it has an average of 19 million daily users (Kaplan, 2023). Not surprisingly therefore it is preferred by many tourists for both keeping in touch with home and for posting travelogues, blogs and photographs.

More recently, however, WeChat has been accompanied and, in some ways, superseded by several other Chinese platforms, notably Little Red Book. Little Red Book (*Xiaohongshu*), or sometimes simply Red, was founded in 2013, as a platform where people share experiences through short videos, graphics and other forms, and is officially committed to 'making the world's best lifestyle at your fingertips'. It is sometimes seen as a combination of Pinterest, Instagram and Amazon. *Xiaohongshu* has achieved niche success in the international social commerce sector of fashion and luxury shopping, reflected in its earlier slogan 'Finding overseas goodies', later changed to 'Inspire Lives'. Users share reviews about experiences in shopping, fashion, travel, and other lifestyle areas, helping others with their purchasing and travel decisions. It was initially a platform and online guide for sharing overseas shopping advice, and thus rather different from WeChat, being centred on beauty and health, and integrating e-commerce. Compared to WeChat and TikTok, Little Red Book is a much smaller social media channel, and its users are skewed towards young women so that it is said to be 'the go-to app for females aged 20–35 years old living in top-tier cities in China' and thus ideal for promoting cosmetics, health, beauty, fitness, tourism, and education. By the mid-2010s it had become very popular and had grown into the world's largest community e-commerce platform. It attracted a mix of internet influencers and celebrities to drive popularity and sales (Song, 2022). These influencers were central to a *wanghong* (网红) economy of social media marketing and online retailing, where *wanghong* (literally 'internet red') refers to 'micro' celebrities and social media influencers, almost all of whom are women (Abidin, 2018; Guan, 2021; Guo, 2022; see Chap. 6). By 2020, Little Red Book had over 300 million registered users (a third of whom were active), with 70% of them born after 1990, typically therefore being members of Generation Z (born between 1995 and 2012) and 90% of whom were females. Most *xiaohongshu* users are metropolitan middle to upper-class women with higher spending power, more than half from 'first tier' cities. *Xiaohongshu* takes video, text, and pictures that record and share lives, lifestyles and experiences.

As GMA, a digital marketing agency based in Shanghai, has stated of the key online users:

> This young and impressionable demographic [Generation Z] is likely to shape society's future values as they become increasingly exposed to new ideas through technology-driven communication tools such as digital media or social networking sites with a lot of user-generated content. This will undoubtedly have an impact on what kind of person we end up becoming in 20 years time! The other [user] groups are cutting-edge white-collar workers, urban trendsetters, singles, exquisite mothers, and hedonists (GMA, 2022)

None of these intriguing groups were described or defined, just as millennials defy easy classification, but GMA had suggestions for those seeking to market products in China: 'try to use numbers, exclamation points, and emoticons with attractive colors (red, e.g.) to let your notes become viral on this social network!' (GMA, 2022). Little Red Book's design is like that of Instagram: three to nine pictures and videos with text descriptions (Chap. 6). The content style is usually very personal, often with lots of emojis, and users can engage with the content through bookmarks, private messages and comment. Proponents of these platforms and their users quickly discovered they could play an important role in tourism.

Social Media and Tourism

Numerous approaches have explored the role of social media—or eWOM—in Chinese tourists' travel decisions (Sun et al., 2015; Wu & Pearce, 2016), as well as in the post-travel stage, when tourists visit websites, WeChat, Little Red Book and other social networks to share their travel notes and photos which then 'help future travel decisions of others' (Wei et al., 2018, p. 2). We explore this usage in Australia in some detail in the chapters that follow. In Australia, as elsewhere, social media users are dominated by Generation Y millennials, generally understood to be people born between 1981 and 1996, and now also by Generation Z Centennials.

Quite simply, social media enable tourists and potential tourists to gain more detail, and consider recommendations, as they consider options, whether in China contemplating travel or at the destination. Even before the rise of smartphones, tourists used social media to organise and share their travel experiences through blogs, online communities and media-sharing sites. The rise of smartphones (see below) has cemented these strategies. Thus Chinese tourists in Thailand have simultaneously made greater use of eWOM to allay concerns about safety, choose routes that would maximise pleasure and confirm the probability of having a satisfying trip, so that they become more independent travellers, relieved of the necessity to be part of package tours (Miao, 2015). Similar usage has contributed to a better understanding of Macao as a destination for enjoying 'heritage sites' and 'shopping and sightseeing' (Qi & Chen, 2019). In Australia, analysis of WeChat was found useful in analysing Chinese tourists' concerns over food choices, language matters, and value for money (Ma et al., 2020). Thus far, such studies have been limited in their capacity to explain the qualitative nuances of social media practice, being modelled with quantitative data sets at a national level. In typically focusing on tourism management themes, culturally inflected social media practices disappear. Hence, we seek to contribute a largely missing qualitative perspective to the burgeoning literature on Chinese social media practices among tourists.

While individuals post information for a variety of reasons, from intrinsic enjoyment, community citizenship and altruism to personal status and egocentricity (Wu & Pearce, 2016; Munar & Jacobsen, 2014; Oliveira et al., 2020), social media posts in tourism contexts often evoke pride at the discovery of something pleasurable and new, celebrating the distinctiveness of places and spaces in what is still an early phase of independent Chinese tourism. It has thus been argued for Chinese tourists that 'sharing on social media creates a 'rosy glow' that leaves senders feeling more positive' (Wong et al., 2020, p. 1773). Here we seek to build on such insights regarding the rationale and motives of social media users, examining the cumulative outcomes for specific sites that are 'marked' through such practices.

Most studies on the relationship between the nature and extent of the use of social media and tourism have focused on the use of social media for planning travel (Fotis et al., 2011; Xiang & Gretzel, 2010) and on the reasons why tourists create and disseminate eWOM content (Bronner & de Hoog, 2011; Munar & Jacobsen, 2014; Yoo & Gretzel, 2011; Wu & Pearce, 2016; Wong et al., 2020). Chinese millennials have indicated that they shared photos on social media for a combination of recognition, status and enjoyment (Li, 2020; see Chap. 6). Any emphasis on Chinese social media also necessitates understanding the relationships between tourism and photography, relationships fostered by various technologies beginning with the rise of the mobile phone (ubiquitous amongst Chinese tourists in Australia) that has influenced the structure and form of Chinese tourism and contemporary tourism in general. Social media have enabled the circulation of images to multiple audiences simultaneously 'in close to real time' (Walsh et al., 2019, p. 1). Since digital cameras are an essential feature of mobile phones, the use of photography to create new tourist places, practices and geographies has become an area of research significance.

This transmission and associated feedback from sympathetic audiences, often equally rapid, has meant that taking photos has become a habitual complement of travel with the camera the main means of experiencing a place 'in aesthetically pleasing and positive ways' (Walsh et al., 2019, p. 1) and, rather than being something developed at a later point to facilitate memories, is increasingly a means of communicating experience and a rationale for visiting particular sites. In Britain in 2017 some 40% of people aged between 18 and 33 believed the most important feature of a travel destination was how good it would look on their social media feed (Guilliatt, 2020). Increased status was at stake. That proportion is likely to be even higher in China where sharing of travel photographs is characteristic of aspirational, youthful middle–class urban residents with particular tastes and lifestyles, and is analogous to the sharing of food (and fashion) photographs (Peng, 2019). Upward social mobility was an anticipated outcome. Millennials (and centennials) are the most likely to use and trust social media (Ek Styvén & Foster, 2018; Oliveira et al., 2020). Over time, the thoughts and opinions of other tourists have become at least as influential a source of information as that coming from travel suppliers (Litvin et al., 2008) and from conventional marketing.

Tourists increasingly prefer and use social media when planning travel because they are easily accessible, up to date, not necessarily with a commercial focus and, by choice, are likely to reflect the perceptions and advice of people similar to themselves. In some cases, social media may be the only readily available sources and often are more detailed than more formal and static sources. Social media have the advantage, compared to such 'older established sources, in enabling tourists to co-produce and share accounts of particular sites, both with and by people already travelling. That may involve filling in empty spaces in an itinerary or modifying the structure of their travels while en route. Additionally, social media are both visual and textual, and the nuances of particular cultures and times (given the evolution of language, symbols and the ability to respond to events) means that particular recommendations and commentaries are often reinforced by others and trusted by fellow citizens because of who they are, what they show and say, and how this is done.

Social media readily enable the targeting of specific messages to particular audiences, further emphasising the role of tourists' imaginations and experiences in contributing to how places and peoples are understood (Urbain, 1989; Hughes, 1992; Crouch et al., 2005). While this was previously the domain of literature, cinema and television, the democratizing power of social media has given voice (or focus) to ordinary people by recognising tourists as not simply passive recipients of information who respond to the agendas set by others in the tourism industry but as active participants and agents (which to varying extents they were prior to social media). The role of user-generated content is likely to expand much further in the future as independent tourism grows in importance.

Social media influence tourists' perceptions of places, but studies that reach such conclusions have often been undertaken in laboratory settings using hypothetical scenarios rather than with actual active tourists (Kim & Fesenmaier, 2017; Magno & Cassa, 2018). Until recently little was known, however, about how the sharing of visual and textual content shapes subsequent tourist imaginaries and experiences, and how it may contribute to creating a new spatiality of tourism, in this case both within China and internationally. The ability of tourists to influence official tourist policy from a 'bottom-up' perspective through the use of social media has been recognised in China, in the 'Blue Tears' of Pingtan Islands where the role of governments 'through planning, policy making and financing' was arguably at least matched by 'social media empowered tourists, who collectively strongly influence or even play a decisive role in the destination image and, through this, the eventual provision of services' (Su et al., 2021, p. 11). Such use has the potential to create a new spatiality of tourism not primarily at the global level but at regional and local scales, situations that we subsequently explore (Chap. 6). This has varied local impacts depending on its intensity, duration, form and the characteristics of the receiving environment, where photographs play a key role.

The Rise of the Smartphone

The emergence of mobile phones, that ultimately became the more technically and sophisticated smartphones with their abundance of functions, including access to the internet, revolutionised many aspects of life, not least tourism. Mobile operating systems enabled them to be used almost anywhere, and few tourists eschewed them. Moreover, smart phones, smaller and simpler than separate digital cameras, and with rapidly evolving camera technology, gradually replaced stand-alone cameras. The first commercial camera phone was released in 1999 and, since 2010, most mobile phones, even the cheapest ones, were being sold with a camera. Camera phones usually had a relatively good lens and high resolution and, in that same year, their sales passed those of cameras which experienced a rapid decline. Camera phones were more compact and convenient and, in any case, almost everyone already had mobile phones for the telephone function.

Over time photographic quality improved, audios and videos could be made and visual zooming between continents followed. As mobile phones are constantly carried, they allowed for capturing scenes and moments at any time, while photos could be sent immediately or posted on the internet, since modern smartphones have almost unlimited connectivity and transfer options. The explosion of personal photography, and especially variants of selfies (self-portraits), enabled people to capture, construct and edit personal memories, and maintain distant social relationships. The new practice resulted in a new way of touristic looking where tourists themselves became the objects of the gaze (Dinhopl & Gretzel, 2016) although the rationale for dissemination was not always different. Selfies offered proof of 'being there' made easier by front-facing cameras. Fascination with selfies originated in Japan in the late twentieth century, with young women engaged in beautifying self-representation, and spread elsewhere in East Asia, including China.

Selfies have been popular on social media, notably on Facebook and Snapchat. That fascination and diffusion has resulted in some criticism that they were usually posted to social media with the intent of getting positive comments and 'likes', reinforcing the notion that the most valuable thing young women had to offer the world was their looks. However others have seen posting selfies as a form of self-expression, as 'conscious attempts to manipulate the self-image' (Liu, 2021, p. 241), or simply to look good in a pleasant setting, so curating and narrating one's own view of desirability, style and achievement, and so be a positive means of actively asserting agency. Multiple photos (and videos) can be taken at a site, then checked and retaken. Performativity is made possible and recorded. Smartphones thus represent 'a historical conjuncture in which notions about identity, individualism, lifestyle and sociality—and their relationship to technology and media practice—require rearticulation' (Hjorth et al., 2012, p. 1). Beyond selfies, smartphone users can be participants in whatever moment and whatever place and event they photograph or video. Photos captured on a camera phone serve to prove the physical presence of the photographer.

Probably nowhere else in the world have smartphones been adopted more rapidly than in China and become deeply integrated into people's everyday lives. Almost every adult in urban China has a smartphone and uses it for multiple purposes, having largely displaced money for most transactions (even for purchasing vegetables from street vendors). Chinese tourists therefore are utterly familiar with smartphones (and such payment systems as WeChat and Alipay) and with creating blogs and vlogs to record their experiences.

By the mid-2010s as many as 78% of Chinese tourists within China were using social media to share their travel experiences after the trip, and half (49%) kept in touch with family and friends while travelling. Moreover, about 45% reported that their own travel was inspired by information posted on social media (Jia, 2017). It seems highly likely that usage was higher outside China, as smartphones have become ubiquitous amongst Chinese tourists in Australia. Alongside the use of social media, that has influenced the structure and form of Chinese tourism and contemporary tourism more generally. Early studies have not surprisingly revealed that their use was affected by the perception of usefulness as they enabled tourists to be more creative and spontaneous, resulting in greater satisfaction (Wang et al., 2016). Since digital cameras are an essential feature of mobile phones, the use of photography to create and document new tourist places, practices and geographies, is an important theme pursued in Chap. 6.

Shopping has always been so important to most Chinese tourists that it cannot be regarded as a niche or as some element of diversity: it is a core element. However its practice casts light on how important mobile phones are in the practicalities of shopping in Australia. Chinese tourists look for products, especially healthcare products (a significant component of purchases), based on images stored in their smartphones, so that 'if a product on site is the same as the image saved on smartphones, Chinese tourists would choose the product without hesitation' (Jin & Cai, 2022, p. 2960). Using smartphones simplifies the requirement to choose the right products and obviates the need to speak English (although the dictionary app on smartphones enables product information, such as the chemicals in health products, to be translated into Chinese). Tourists were also able to assess the value, quality and price of additional products by searching for them on Taobao (a Chinese mobile app for online shopping). Price advantages are a primary reason for shopping in Australia, as elsewhere (Zhu et al., 2016), and phones enable instant comparison of prices on e-commerce platforms at home. Barcodes are scanned in, and conversion rates undertaken using the calculator on the smartphone. Photographs of products, such as clothes and shoes, can be sent instantly, enabling intended recipients in China to choose which they prefer. Shopping becomes a performance, and such practices enable tourists to overcome 'information asymmetries' otherwise typical of most tourist shopping (Jin & Cai, 2022). Equally significantly, as we discuss later, these practices have important analogies with recommending and choosing the places of tourism.

Emerging Diversity?

Here we trace the rise of diversity and the emergence of niches, partly associated with social media and greater individualism, and partly with a relative decline of group tourism. We assess how this further evolved and what constraints it faced, during a relatively early era of social media and photography. In Chaps. 5 and 6, we take these themes much further as photography became of significance in the identification of new sites, exemplified in the newly emergent niche that combined education or campus tourism and the search for traces of the mythical figure of Harry Potter in the main building of the University of Sydney, neither of which had been anticipated or planned (Chap. 5). That eventually resulted in 'marking' places previously absent from or marginal to established tourist circuits.

As Chinese tourism grew in numbers and in the diversity of visitors, the tourism experience changed from more institutionalised and standardised package tours, that nonetheless continued to characterise tourism, and indeed still characterise tourism within China (China Tourism Academy, 2020), to more independent travel. With return visits, increased student numbers in Australia (linking to VFR tourism), greater experience, greater expectations, younger tourists (with some English language ability) and longer stays, tourism aspirations gradually changed. At the same time social media enabled flexibility within Australia, and tourists sought to use them to demonstrate that they had been somewhere different. Our own interviews with package tourists at Sydney University (Chap. 5) revealed that several were uninspired and uninterested there and would have preferred something and somewhere else. Injecting their own choices gradually became feasible. Rather simplistically, as the general manager of UnionPay, China's dominant card provider, suggested (no doubt with tongue firmly in cheek): 'There is a saying that Chinese tourists get off the tourist bus just to take pictures and go the bathroom, that is no longer the case' (quoted in Tan, 2017). Likewise, a travel company director has noted:

> Today's Chinese traveller is much younger, savvy and inquisitive and much more confident about what they do and where they go. Only a few years ago a trip to the beach for many Chinese tourists would have involved just dipping their toes in the water and taking photos. Now they are queuing up for surfing lessons (Tourism Australia's managing director, quoted in SMH, 13 May 2017).

As tourism became younger, more individualistic and more personal, it also became much more participatory. Photographs were no longer enough.

The growth of the Chinese population of Australia, and especially of students, brought a surge in Visiting Friends and Relatives (VFR) tourists, who make up around a quarter of Chinese tourists. Over much the same time period, a rise in Japanese student numbers had belatedly resulted in a parallel increase in VFR tourism (Kashiwagi et al., 2018). VFR tourism is part of a broader sector where 31% of Chinese visitors are part of a wider socio-economic system with connections to relatives, but also to education, business, property investment and commerce (Austrade, 2017; LEK, 2017). Around a third of all tourists have at least secondary objectives beyond the notion of normative tourism. Since the Chinese population in Australia was now growing extremely quickly through migration (although COVID-19 later curtailed that growth) and there were parallel growing numbers of temporary Chinese students in Australia (166,000 in early 2018), both increased familiarity with Australia and increased resident numbers were factors in stimulating tourism and family reunions (Austrade, 2017). The international education market in Australia has long been strong, with Australian universities highly regarded in international academic comparisons, and tourism offering an opportunity for some to examine and compare these facilities (Xia et al., 2018; Chap. 5), a shift away from more obvious forms of tourism.

A further distinct group, sometimes seen as one component of the tourism sector, but rarely addressed in the same marketing terms—since they were outside the system in many ways—is that of working holiday makers (WHMs). Much like students, they tended to be younger people, largely because this visa option is generally limited to people in the age range 18–30 years (Australian Government, Department of Home Affairs, 2022). Chinese WHMs had to have tertiary education and English skills, so that students and WHMs often had the same ages and interests and students sometimes worked in the same places as WHMs. Working holidays have allowed youth to travel to Australia from China (like many other countries) to undertake paid employment but also spend time travelling; many worked in agriculture (which often enabled them to extend their visas) or in the tourism and hospitality industries (which were habitually short of workers, but popular places for short-term work) and often stayed for two or three years. China entered the programme in 2004 (and Hong Kong and Taiwan joined separately).

Although Australia enabled WHMs from Asia, including up to 5000 Chinese a year, between the ages of 18 and 30 to work and travel in Australia, actual numbers were much less. Despite their primary commitment to work most came for the same sort of tourist experience as that of other younger Chinese, and their status, rationale and experiences were much the same as those of WHMs from Taiwan (Lin et al., 2012; Wang & Connell, 2021). A sample of 15 Chinese WHMs in the mid-2010s revealed that all had university degrees (as required), many held previous white-collar employment (such as teacher or IT developer), so being similar to other groups of younger tourists, and most anticipated at least some degree of cultural exchange. Most had come, firstly, to exert a degree of independence, freedom and autonomy from families (and perhaps from society as a whole); secondly, to acquire knowledge and skills that might later be beneficial (such as English language skills); thirdly, for pleasure, escape and relaxation (despite employment being at the core) and, fourthly, for 'ego enhancement', in an activity that one suggested 'only cool guys will do'. Most were enthusiastic about putting their own pictures and experiences online. As one WHM said, 'It's so cool to see all the pictures and experience-sharing online'. Another noted:

> I got a friend who did a working holiday last year. He posted a lot of fancy pictures about his life and travel on WeChat and got over 1000 Likes. I want to have the same experience ... this is a chance to show that you are different (quoted in Fang et al., 2021, p. 9)

Taking photographs motivated all WHMs, and though they increasingly followed in the same footsteps as earlier workers, they sought to distinguish themselves and gain status by circulating unique or unusual photographs. This matched a wider context where people felt pressured to compete, compare material possessions and share their experiences on social media in order to impress.

Beyond the existence of the WHM system (and the ability to earn money through this that would finance tourism), the reasons for coming to Australia and their expectations were much like those of tourists. Firstly, the range of natural scenery and tourist attractions was popular, so seeing wildlife and visiting such 'landmark attractions' as the Great Barrier Reef and Uluru were important. Apart from specific attractions, Australia's blue sky, fresh air, sunshine and beaches were all attractive, echoing a desire for 'escape and relaxation'. Many expected to undertake activities, rare in

China, including self-driving holidays, bungee jumping and sky diving: novel experiences that would show their peers that they were unique (Fang et al., 2021). Safety was a final Australian pull factor. Most had carefully undertaken research through social media and contacted other former WHMs. These pre-occupations and interests 'aligned with prior studies that the young generation in China are keen observers of social trends, seek information from social media, and trust the opinions of their peers' (Fang et al., 2021, p. 10; see also Wu & Pearce, 2016). Although they expected some degree of adventure, their aspirations were little different from those who had preceded them. Ultimately what both Chinese students and WHMs anticipated was not substantially different from that of short-term tourists but their longer presence meant that they had much greater expectations of more cultural understanding and language learning along with some acquisition of skills. As one WHM suggested, 'I hope to stay in a local community, just like a local, so that I can experience the real Australia' (quoted in Fang et al., 2021, p. 11). That at least suggested a determination to engage more fully with Australia, a possibility not really open to most tourists because of the short duration of their visit.

Despite the growth of VFR tourism and the partial tourism of WHMs, some key aspirations and elements never changed. At the core of tourism for many, and effectively unchanging, has been shopping. Rather more than for other national groups, shopping is a significant element of tourism. While Europe has advantages through its fashion labels, shopping in Australia has centred on items perceived as 'clean and green'. Products popular with Chinese tourists include baby milk formula, dietary supplements and vitamins, alongside wine, honey, clothing and electronic products. By 2016, a quarter of all expenditure was estimated to be on skincare items, health products and gifts; soon afterwards, AuMake became the leading platform utilising Chinese influencer services to promote Australian (and New Zealand) brand products, including health and beauty products, pharmaceuticals, alcohol and 'mum and bub' products. Trust and safety have been vital for items that are ingested, especially since most couples had only one child. Companies such as the Chemist Warehouse chain benefited considerably. Chinese tourist expenditure was therefore particularly high, compared with other national groups, since some part of tourism (often in the evenings) was devoted to strategic purchasing of food and medical items, in addition to the usual range of tourist purchases (often associated with local nature, such as Ugg boots and kangaroo skin leather goods). That also influenced local businesses, whether or not

Chinese owned, and whether or not in the tourist industry, and also the national and state government, to see virtues in developing particular marketing strategies.

Chinese businesses in Australia, especially in and around Sydney's Chinatown, expanded to take advantage of the growing significance of the *daigou* trade, and the growing participation of tourists in *daigou* to the extent that it could be considered almost as a distinct niche in itself, within the wider context of shopping. *Daigou* (代购), literally buying on behalf of someone else, usually involved Chinese residents (often students) and tourists buying goods to be shipped or taken back. One estimate suggested that in 2017 there might be as many as 40,000 *daigou* traders sending Australian goods to China, by which time AuMake had six stores in Sydney alone, and there were an estimated 1500 linked stores in Australia (Sier, 2017; Kohler, 2017). Such goods often included foods (notably infant formula), vitamins and cosmetic products. Niche tourism or not, it was concentrated in the largest metropolises, where many small Chinese trading businesses were concentrated (Tham et al., 2022). Incidentally, in several cases, it blurred boundaries between tourism as leisure and tourism as business.

Nature and wildlife, always at the centre of Chinese tourism, have continued to dominate the rationale and thus destinations for travel. Preference for clean environments, blue skies and distinctive animals, an environment that is managed and manageable, and thus perceived as being safe, and in harmony with nature, gave places like the Great Barrier Reef, a comparative advantage, but one that was diluted by distance (Zhang & Peng, 2014; Xia et al., 2018). Nature tourism subsequently spread beyond the east coast. In the distinctive context of the Great Ocean Road, where two-thirds of the Chinese tourists were independent travellers, nature, wildlife and the scenic driving experiences most enthused visitors, and especially at the iconic Twelve Apostles (Pearce et al., 2015; Han & Cheer, 2018). Distinctive animals are factors, and account for relatively remote sites seeking to attract Chinese tourists through animals. Dubbo has advertised the Western Plains Zoo, Uluru promoted camels, alongside the rock, and Melbourne offered Phillip Island penguins. Featherdale Wildlife Park, in the western suburbs of Sydney, while not specifically identifying Chinese tourists as its marketing target, promotes on its website its geographical location, the native and iconic species and the 'personal experience' awaiting tourists, between two iconic destinations. Its website encourages visitors to 'have a personal encounter with a Koala and hand feed Kangaroos and Wallabies. Located at Doonside, en route to the famous Blue Mountains, Featherdale is Sydney's 'hands-on' wildlife experience'. Likewise, the website of Ballarat Wildlife Park (in mid-2023) promoted the ability to see 'over 100 free-roaming kangaroos (you can hand-feed), more than 35 koalas, wombats, emus', and it depicted a smiling Chinese visitor cuddling a koala. This aligns with what have hitherto been city-focused and nature-based experiences. Similar attractions in Brisbane are 'suburban or peri-urban areas such as farms and wildlife parks ... all of which were less than 60-minute visits, but all of which included shopping'. It took time for tourists to venture further afield. Indeed the itineraries of large group tour providers indicate that very little time is spent outside major metropolitan areas, and then only as fractions of a single day to the Blue Mountains, the Great Ocean Road or Phillip Island (Weiler & Chen, 2016, p. 252), contributing little if anything to the economy of these places, but creating an urban periphery of 'wildlife' parks.

In many respects, this urban periphery was as far as many tourists ventured. Exactly which places Chinese tourists visit is impossible to determine; but surveying the origin of the posts enables a reasonable approximation (although here too there is a metropolitan bias). Surveying more than 51,000 Weibo posts in 2018 revealed that 41% went from Sydney closely followed by 37% from Melbourne. Brisbane was a distant third with 8% and Cairns even further behind (4%).All other cities, including Canberra and other state capitals, recorded trivial numbers (Chen et al., 2020). A supposed passion for selfies with the famous quokkas of Rottnest Island (off Perth) (McHugh, 2017) was not enough to draw many tourists to the west. The posts also indicated that many tourists only travelled in a short loop or path between Melbourne and Sydney and, much less frequently, included Brisbane. Sample surveys of FIT travellers, more likely to be interested in the regions, but deterred by distance, time and fear of safety, further pointed to the pronounced metropolitan focus (Austrade, 2019). The periphery of metropolitan Sydney and Melbourne was as regional as Chinese tourists usually experienced.

Beyond established wildlife, nature and icon tourism (that gave east coast cities an advantage), new foci and greater independence did emerge. Over time, and with more time, museums and galleries steadily became more popular with some shift of emphasis from natural wonders to cultural attractions. The Sydney Opera House has led the way. Chinese visitors were twice as likely to cite it as a reason to visit Sydney as British tourists, and were even more likely to attend events there, because it was perceived as the only 'high status' cultural possibility in the country (Gibson, 2018). While the Opera House was the epicentre of tourism, the regions were largely unknown and so unvisited, and tourists preferred to see animals in controlled environments rather than in the wild. Language, unfamiliar culture and nature, the lure and pressure of metropolitan shopping, and absent or limited Wi-Fi access (including the inability to pay through Alipay and UnionPay), conspired to confine most tourists to occasional controlled forays to the fringes of the cities, and security contributed to involvement in group tours. Indeed tourists were pleasantly surprised by their discovery of safety and security at the Great Ocean Road (Han & Cheer, 2018) while discouraged from visiting Port Arthur (Tasmania) by its poor Wi-Fi connectivity (Carson & Pennings, 2020). Not all would have welcomed greater independence but the aspirations of those that did are well summarised in the words of one middle-aged woman:

> I know that the quality of group tour is not so good—however the reason I still chose it is that I can't drive on the left-hand side and more importantly I can't speak or read English. If there is a small group tour of only 3 to 5 people with a high quality tour guide that would be awesome (quoted in Han & Cheer, 2018, p. 175)

Just a handful of regional towns have benefited, notably Ballarat, a hundred kilometres from Melbourne, with both the relics of nineteenth century gold mining and a wildlife park. By contrast, Colac a smaller town not far from the Great Ocean Road (where its Chamber of Commerce claimed that 2000 Chinese visitors passed through annually), enthusiastically adopted QR code technology in places likely to be frequented by tourists, and advertised in Mandarin on storefronts, but the numbers scanning the codes increased from 8 to 69 over a period of three months in 2018, while the Chamber observed that 'Local store owners were apprehensive about engaging Chinese customers. Communication was the great barrier' (quoted in Victoria The Education State, 2018). It was perhaps unsurprising that this disappointing outcome was noted under the heading 'The Great Wall'. In New South Wales, Mudgee was the centrepiece of a state government advertising campaign in Hong Kong, oriented to wine, clean air and the Western Plains Zoo in Dubbo, well summarised in the title of a media account: 'NSW invites armies of rich Asian tourists to Mudgee and Dubbo' (*Daily Telegraph*, 8 May 2017), but it had no discernible impact and failed to be extended to the mainland as had been intended. Few other towns, even those near tourist routes, attempted to enter the market.

Other parts of Australia have scarcely benefited, although the will was there; as one tourist of independent means explained:

> The Australian cuisine, weather and the space is what is really attractive for me... a real experience rather than just shopping. Tasmania's nature really attracts me. I want to visit the farms, vineyards, markets etc. Hopefully I can see the Southern Lights too (A 38-year-old in Tasmania, quoted in Smith, 2016, p. 42).

Without direct air connections, Tasmania marketed itself to appeal to younger, more cosmopolitan and bilingual Chinese tourists, willing to experience more non-Chinese food. As noted earlier, the Northern Territory sought affluent 'niche and high-end travellers' rather than the 'mass market bus travellers' (Chap. 3). Nonetheless, even such varied approaches had limited outcomes—spatially or by niche—in this case constrained by an absence of direct flights. Distant places might have resigned themselves to disappointment: Australian metropolitan residents are themselves conservative in tourism habits with the 'sandstone curtain' of the Great Barrier Range seemingly discouraging inland tourism by many who live on the east coast.

Despite a degree of diversification, familiar iconic places retained their primacy, and were almost essential to first-time visitors. Sydney Harbour and the Great Barrier Reef remain major destinations, while metropolitan bias is accentuated by a continued interest in shopping. However, popular locations that combine beaches, blue skies, clean environments and perceptions of safety include Hayman and Hamilton Islands, where there are Chinese owned hotels, and also Sydney. Non-metropolitan tourist destinations include wine regions such as the Yarra Valley (near Melbourne) and the Hunter Valley (near Sydney), which increased in popularity as Chinese alcohol consumption patterns evolved. Red wine proved particularly attractive in China (Gray, 2018) but the thirst for red wine extended into wine tourism only where production regions were scenic and readily accessible from metropolitan areas. Language barriers, lack of information, and transportation curtailed the extent of wine tourism (Gu & Huang, 2019).

More generally, language barriers, food preferences and other factors still disadvantaged regional and more individualistic tourism, despite tourist groups becoming smaller and more specialised. Significantly one winery in the Adelaide Hills began generating unusual interest and significant sales in China after employing a Chinese-speaking student who could converse with the tourists (Cai, 2017), being one small example of the role of the diaspora in the tourism industry. Otherwise tourism in the regions offered scope for lyrical and bucolic marketing photographs of happy Chinese visitors, usually with either kangaroos or koalas, but it recorded exciting possibilities rather than substantial numbers.

A Hint of Difference

Not everything was passive. Adventure tourism expanded but remained unusual (Liu et al., 2017). Shooting, golf, diving and even skydiving became popular with a few, but serious physical activities were relatively uncommon. Indeed fear permeated many tourists' attitudes to most unusual activities unavailable in China; even 'seeing Australian animals' was an exciting activity, visiting wineries was 'thrilling and exciting' (Gu & Huang, 2019, p. 215) while 'snorkelling proved very challenging' (Jin & Sparks, 2017, pp. 209, 211). The same concerns related to canoeing and rafting, where Australia offered faster flowing rivers, more powerful rapids and greater personal responsibility for safety, all of which made venturing onto rivers more dangerous for the tourists, who were more used to being passive passengers on calmer waters (Buckley, 2014; Jin & Sparks, 2017). More subtly, even Generation Y visitors, many of whom were the only child in their family, felt they needed to be safe and not engage in anything dangerous that would risk their health or life, an attitude that conflicted with the marketing of thrills and adventure, whilst some females were fearful of becoming tanned because of sun exposure (Gardiner & Kwek, 2017, p. 501). Sun exposure discouraged some from surfing but a greater disincentive was that many tourists had never seen the sea, could not swim and were quite apprehensive about it (Murphy, 2017) as significant surf was rare on China's coastline.

Surfing was however prominent amongst the more active activities, eventually becoming popular amongst more youthful tourists (Gardiner, 2021), and lending itself to photogenic marketing. The relatively small number of male influencers tended to post photos of surfing (Chap. 6). It was inevitably an accessible coastal activity. Adventure tourism was more closely related to international student activities, with Gardiner and Kwek (2017, p. 496) observing that, in 2014, many of these thousands of students in Australia 'want to experience the freedom of international travel that enables self-development and maturity, but they also have a strong sense of familial obligation and distinctive cultural beliefs that underpin their consumer decision making'. In many senses these adventure tourists are much like Western backpackers, but the Chinese international 'students tend to take shorter trips and seek less adventurous and more traditional tourism experiences than backpacker travelers, such as sightseeing and shopping' (Gardiner & Kwek, 2017, p. 496). Once again, Chinese students and tourists were urban and conservative in their choices. Some optimism was attached to the emergence of a Chinese market for winter snow tourism, with Chinese urged nationally to take up winter sports ahead of the 2022 Winter Olympics in China (Bagshaw, 2018) but COVID-19 prevented that niche from emerging in the short-term, although it retains potential (and would spread tourism over another season).

The ability to travel more widely was also inhibited by concerns over safety, and driving on the 'wrong' side of the road initially discouraged car hire, especially of large off-road recreational vehicles (RV), but the combination of flexibility of travel and the greater ability to experience nature eventually stimulated slow growth. Those who engaged in RV tourism were relatively young (in their twenties or thirties), came from large cities, were wealthy (enabling vehicle hire and independent hotels, and so spent more), and had a longer time to be in Australia (Wu & Pearce, 2014; Wu, 2015). Several linked up with international students which reduced language problems. Many were return visitors, and/or had travelled abroad previously. Most of these characteristics are likely to characterise other travellers claiming particular niches. Perhaps ironically, most of those who had RV holidays, focused on the east coast extending as far south as the Great Ocean Road, and centred on the Brisbane-Sydney stretch (Wu, 2015) within a limited number of recommended routes.

Another niche growth area was weddings. Australian scenery provided a distinctive and desirable backdrop for photographs and memories (while fewer people are likely to attend, and it is therefore less expensive). One Chinese wedding planner from Melbourne observed:

> Australia's amazing natural sceneries, multi-cultural and friendly environment are the key factors that attract Chinese couples. Whether it is a beach wedding, or a church wedding, or a winery wedding, the variety of locations Australia possesses will help the couple to realize their dream of the perfect unique wedding (ABC Radio, 21 January 2015)

It was not necessary to actually be married in Australia, since that could be complicated, and some couples brought their wedding dresses and suits to change into at the appropriate aesthetic photographic moment. Interest in wedding photographs was part of a wider interest in colourful backdrops, and placing oneself in the photograph, which could partly be satisfied in established destinations from the Great Ocean Road to the Great Barrier Reef and even, occasionally, extending into colourful Australian deserts. However new popular destinations could unexpectedly emerge for unusual reasons. One such destination was Tasmania's Bridestowe lavender fields (Fig. 4.1); another proved to be Sydney University (Chap. 5)—both elements of a growing diversity.

Tourism in Tasmania was boosted by demand to see the 'homeland' of Bobbie the Bear, a hand-stitched, scented teddy bear produced by Bridestowe Lavender. Initially created as a solution to selling surplus lavender, Bobbie the Bear has driven a mini tourism boom since 2013, after he appeared on Chinese social media, being hugged by the famous model Zhang Xinyu (and other actresses) in a much-circulated selfie. A year later President Xi was presented with his own bear. In that year some 65,000 tourists visited the Bridestowe lavender farm and 60% of its revenue came

Fig. 4.1 Bridestowe

from Chinese visitors. Demand was so high that a quota of one bear purchase per visitor had to be enforced (ABC News, 9 October 2014), a decision that 'brought meltdowns in the store' as tourists could only obtain one Bobbie, described by the Bridestowe manager as 'boutique, unique, authentic and very personal' (LEK, 2017, p. 26). Wedding photos were frequently taken with the lavender farm as a backdrop. Quite simply, as the husband in one couple, posing in wedding costumes a year after the event, said: 'because the lavender farm is very beautiful and my wife likes purple'. Meanwhile, the Bridestowe manager enthused over the outcome of several years of targeting China, noted that purple was equated with romance and envisaged that the farm would become one of the 'hotspots' for wedding photographs alongside Uluru and the Great Ocean Road (quoted in Hooper, 2017). The outcomes proved vivid demonstrations of the importance of photography and colour in culture and tourism, while Bobbie the Bear was enlisted in a state road safety campaign aimed at Chinese tourists. The scope for photography and the role of influencers (Chap. 6) were beginning to create a new and distinctive tourist geography.

Continuity or Change?

Despite substantially increased tourist numbers through the twenty-first century, the dramatic growth of social media, more return tourism and expanded WHM and student numbers, and repeated statements and arguments that independent tourism was becoming more significant, there was only limited evidence that this was so (at least until the late 2010s). For most tourists, the structure of tourism—usually confined to a two week break in China (and thus somewhat less time in Australia)—offered limited scope for flexibility and escaping the 'norm', especially for first time visitors. Language barriers, food preferences, Wi-Fi access and other factors disadvantaged regional and more individualist tourism. Tourists stressed safety and security and that came with numbers. None would have contradicted the CEO of Tourism Australia when he concluded: 'In all our research, it tells us safety and security is [sic] the top sort of driver of choosing Australia' (quoted in Koob, 2017). Where shopping was everywhere an important component of tourism, and Chinese were significant spenders, local metropolitan tourism operators had no incentive to encourage tourism to the regions. The superstructure of the industry was firmly metropolitan. Urban entrepreneurs, inertia and a metropolitan bias were entrenched.

Alongside the emerging possibilities of social media, smartphones retained a functional value and Chinese tourism continued to centre on Australia's east coast, iconic sites, and a preference for suitably controlled nature, blue skies and clean air (almost a mantra), and shopping. Tourism remained primarily metropolitan; package tours were brief, spatially limited, and centred on Australian cities with occasional forays beyond. These core elements never lost popularity, despite satisfaction with such tours being less than with independent travel, partly because they required fewer or no English language skills. Familiar Chinese food and shopping opportunities were at least as welcome. Indeed, until recently, language issues were the greatest concern of Chinese visitors (Jin et al., 2016; Han & Cheer, 2018). Yet all surveys of aspirations point to tourists wanting to experience rather more. Even by 2005 it was clear that some tourists were frustrated by what they saw and what they failed to see and that they wanted a more flexible itinerary. As one stated, 'My dream was to visit a casino, sing and dance with local people, hear a concert or opera and see a western movie but my dream was not fulfilled' (quoted in Weiler & Yu, 2006, p. 18). However optimistic that might have been, time and lack of information prevented its fruition, as others found limitations to their own expectations. With little time, little sense of what might be available and embedded in a tour, diversity and difference were impossible.

While numerous reports pointed to the emergence of greater diversity, distinct niches and segments, and some spread of tourism from the east coast (helped by belated direct flights from China to Western Australia), no quantitative data record such shifts, or even what tourists did over a two-week period or where they went. Tourist posts, however, are a proxy and reveal the massive metropolitan dominance of travel (Chen et al., 2020). By contrast, even seemingly rigorous studies of actual tourism lacked methodologies (e.g. Austrade, 2019). Media articles often ignored any distinctions between tourists from the PRC, Hong Kong or Taiwan. Typically academic literature (and the optimism of marketing campaigns) suggested that tourists were becoming more sophisticated in their travel behaviour and were seeking more active and unusual experiences, but there was little data to support that assumption, although some studies were happy to discover and distinguish distinctive tourist segments: for example differentiating those seeking 'imitation', 'compensation psyche' or 'esteem seeking' (Austrade, 2019) without stipulating who they were (or what that might mean, in a cross-cultural context). It is simply impossible to tell how many tourists chose to climb the Harbour Bridge rather than simply observe it from afar. Wealthier and more educated tourists were more likely to opt out of group tours, but how many travelled outside the metropolises, for how long and who they were, is almost entirely unknown. Little was known about which tourists sought out what activities and practices, why they chose particular tours and how they experienced them. Gender and age distinctions are quite elusive. Authenticity has been argued to be important for many travellers, in the sense of some degree of integration of physical building, spiritual meaning and the site itself, in a relatively 'natural' and authentic setting (Xu et al., 2008), but such a search for harmony may no longer be true. Who chose what and why and how much agency and volition tourists had at different points in their travels is equally unknown. This lacuna was one factor in our undertaking our own study, which we describe in the following chapters.

Typical media headlines pointed to new trends and developments (e.g. 'New Chinese tourist trend creating opportunities for Australian businesses', *Business Asia Hub*, 18 March 2017) but most seemed anticipatory rather than actual. Marketing was becoming more adventurous and optimistic, but a different future was always about to happen. As Ross Beardsell, Executive Vice-President of the commercial real estate group, JLL, observed in 2019: 'We're yet to even touch the surface of the Chinese luxury traveller. Australia does a superb job attracting the group tour segment from China, but we're yet to tap into the luxury market—the guest that will stay in 5-star hotels, referred to in China as 6-star hotels (JLL, 2019). That JLL were significantly involved in Chinese tourism issues did indicate that at least one niche was emerging, where tourism was linked to the search for property.

Nonetheless, some Chinese tourists were avoiding group tours and coming with friends and family members only, often linking up in Australia with students and temporary workers. Despite the lack of formal data on independent tourists, a myriad of qualitative observations and other skeletal data indicated that the numbers and proportion of FIT travellers were increasing, and that they were younger, better educated, more affluent, mainly from the first-tier cities, and stayed longer than others. Time and money prevented their travelling more into the regions. Two-thirds of them benefited from eWOM (Austrade, 2019). Some engaged in more specialised interests, often involving more hands-on activities (whether surfing or self-drive tourism or wine tasting). VFR tourism grew in popularity as the population of Chinese-Australian citizens and students grew. Return visiting was more common, albeit partly a function of business and education visitors, some of whom came simply for conferences or conventions, or seeking real estate (so pointing to the need to clarify definitional issues over who are tourists) and because of VFR tourism. Greater familiarity has contributed to more distinctive and return visits. Repeat visitors stayed longer, spent more, were unlikely to wish to see only the same things again, and were more confident of travelling independently. Associated with geographical expansion and smaller groups has come demand for more diverse experiences, such as wine tours and farm visits, and more active, corporeal experiences, as younger tourists were more confident about independent travel, however much they were a still conservative spearhead.

Despite more adventurous marketing, it struggled to make an impact (Chap. 3). Thus the *NSW China Tourism Strategy 2012–2020* of 2012 sought to 'build consumer awareness' of various experiences: 'Food and Wine, Self Drive, Golf, Resort Stays, Soft Adventure, Luxury, Iconic Sites and Events' but it was ahead of its time. Even where such activities were taken up, they were usually simply adjuncts tacked-on to group tours that were linked to 'standard' tourist activities and place. By contrast, the remarkable and unanticipated success of colourful Bridestowe (and, as we discuss in the next chapter, the University of Sydney) both happened 'by accident' in the sense that neither were marketed to Chinese tourists. Equally, real estate tourism and education (or campus) tourism, insofar as they were tourist niches, were never promoted. Likewise, tourists who participated in any of these activities benefited rather more from social media than previous tourist cohorts.

In the next chapter we turn to an unusual 'dual' niche—education and Harry Potter—centred on Sydney, so still quintessentially metropolitan (and east coast), where social media and smartphones took tourism well beyond old-fashioned guidebooks and beyond the ambit of marketing. Photography became of significance in the identification of new sites, even more directly linked to the social media practices of Chinese tourists, and the power of influencers, that 'marked' places previously absent from or marginal to established tourist circuits, as in the case of the newly emergent niche linking campus tourism with the search for traces of the mythical figure of Harry Potter in the buildings of the University of Sydney (Chap. 5). We then explore how the use of social media, and particularly Little Red Book (*xiaohongshu*), has created a subtle and unanticipated new touristic geography of Sydney (Chap. 6). This dual niche more evidently responded to matters of the heart. Nonetheless, the 'new tourism' never replaced group tours, just as the addition of new sites never displaced previous marked places and icons (albeit creating new ways of seeing such places). Even as tourism was tentatively involving a degree of diversity, and interest in the regions, it was taking a new and intriguing turn at the metropolitan core.

Bibliography

Abidin, C. (2016). Visibility labour: Engaging with influencers' fashion brands and #OOTD advertorial campaigns on Instagram. *Media International Australia, 161*(1), 86–100.

Abidin, C. (2018). *Internet celebrity: Understanding fame online.* Emerald.

Austrade. (2017). *Visiting friends and relatives in Victoria: The China and Indian markets.* Austrade.

Austrade. (2019). *Chinese free and independent travellers – Their potential for regional Australia.* Austrade.

Australian Government, Department of Home Affairs. (2022). *Working holiday maker visa program report.* Australian Government, Department of Home Affairs.

Bagshaw, E. (2018, August 5). Ski fields fake it till they make it. *Sydney Morning Herald*, 10.

Becken, S., & Scott, N. (2016, July 8) How Australia can capitalise on Chinese tourism. *The Conversation.*

Bronner, F., & de Hoog, R. (2011). Vacationers and eWOM: Who posts and why, where, and what? *Journal of Travel Research, 50*(1), 15–26.

Buckley, R. (2014, April 25). Ecotourism for the Chinese: Up the creek without a paddle. *The Conversation.*

Cai, P. (2017, January 31). Australia must capitalize on Chinese diaspora. *East Asia Forum.*

Carson, S., & Pennings, M. (2020). Online and on tour. The smartphone effect in transmedia contexts. In M. Mansson, A. Buchmann, C. Cassinger, & L. Eskilsson (Eds.), *The Routledge companion to media and tourism* (pp. 414–423). Routledge.

Chen, J., Becken, S., & Stantic, B. (2020). *Identifying Chinese tourists' travel patterns in Australia from Weibo posts.* Unpublished paper to CAUTHE conference, Auckland.

China Tourism Academy. (2020). *2019 annual report on China's outbound tourism development, Beijing.* [in Chinese]. Accessed May 3, 2023, from http://eng.ctaweb.org.cn/ctaen/c09/202103/9189e906cd044e0ebcd388693e7fb6ff.shtml

Chiu, C., Ip, C., & Silverman, A. (2012). Understanding social media in China. *McKinsey Quarterly, 2*, 78–81.

Chu, S., Deng, T., & Cheng, H. (2020). The role of social media advertising in hospitality, tourism and travel: A literature review and research agenda. *International Journal of Contemporary Hospitality Management, 32*(11), 3419–3438.

Chung, J., & Buhalis, D. (2008). Web 2.0: A study of online travel community. In P. O'Connor, W. Höpken, & U. Gretzel (Eds.), *Information and communication technologies in tourism* (pp. 70–81). Springer-Wien.

Crouch, D., Jackson, R., & Thompson, F. (Eds.). (2005). *The media and the tourist imagination: Converging cultures*. Routledge.

Darragh, M. (2009). A whole new world: Companies are learning how to reach Chinese consumers cost-effectively via social media tools. *Communication World, 26*(6), 35.

Dinhopl, A., & Gretzel, U. (2016). Selfie-taking as touristic looking. *Annals of Tourism Research, 57*, 126–139.

Ek Styvén, M., & Foster, T. (2018). Who am I if you can't see me? The 'self' of young travellers as drivers of eWOM in social media. *Journal of Tourism Futures, 4*(1), 80–92.

Ettlinger, N. (2019). Digital presumption, crowdsourcing, and the digital regime of work. *New Formations, 99*, 52–65.

Fang, Y., Chien, P., & Walters, G. (2021). Understanding the emerging Chinese Working Holiday market. *Journal of Vacation Marketing, 27*(1), 3–16.

Fotis, J., Buhalis, D., & Rossides, M. (2011). Social media impact on holiday travel planning: The case of the Russian and the FSU Markets. *International Journal of Online Marketing, 1*(4), 1–19.

Gardiner, S. (2021). Chinese learn to surf: Learning ambidexterity and tourism experience innovation. *Journal of China Tourism Research, 17*(1), 142–162.

Gardiner, S., & Kwek, A. (2017). Chinese participation in adventure tourism: A study of Generation Y International Students' perceptions. *Journal of Travel Research, 56*(4), 496–506.

Gibson, C. (2018). Tourism nation: Changing meanings of Cultural Tourism. In G. Turner, D. Rowe, & E. Waterton (Eds.), *Making culture. Commercialisation, transnationalism, and the state of 'nationing' in contemporary Australia* (pp. 116–128). Routledge.

GMA. (2022). *How to get started on the Chinese App Little Red Book aka RED as a business?* https://marketingtochina.com/chinese-app-little-red-book/

Gray, D. (2018, July 25). China's demand is a corker for wine as records tumble. *Sydney Morning Herald*, 20–21.

Gu, Q., & Huang, S. (2019). Profiling Chinese wine tourism constraints: A comparison of Chinese Australians and long-haul Chinese tourists in Australia. *International Journal of Tourism Research, 21*(2), 206–220.

Guan, Z. (2021). Chinese beauty bloggers: Amateurs, entrepreneurs, and platform labour. *Celebrity Studies, 12*(2), 326–332.

Guilliatt, R. (2020, January 25). Invasion of the insta tourists. *Weekend Australian Magazine*, 20–23.

Guo, J. (2022). The postfeminist entrepreneurial self and the platformisation of labour: A case study of *yesheng* female lifestyle bloggers on Xiaohongshu. *Global Media and China, 7*(3), 303–318.

Haiyan, C. (2010). An Impact of social media on online travel information search in China. In *3rd international conference on information management, innovation management and industrial engineering* (Vol. 3, pp. 509–512).

Han, X., & Cheer, J. (2018). Chinese tourist mobilities and destination resilience: Regional tourism perspectives. *Asian Journal of Tourism Research, 3*(1), 159–187.

Harwit, E. (2017). WeChat: Social and political development of China's dominant messaging app. *Chinese Journal of Communication, 10*(3), 312–327.

Hjorth, L., Burgess, J., & Richardson, I. (2012). *Studying mobile media: Cultural technologies, mobile communication, and the iphone*. Routledge.

Hooper, F. (2017, January 11). Chinese newlyweds make the most of Tasmania's blooming lavender. *ABC Northern Tasmania*.

Hughes, G. (1992). Tourism and the geographical imagination. *Leisure Studies, 11*(1), 31–42.

Jia, Y. (2017). A review of tourist behaviour studies on social media, based on overseas literature. *Tourism Tribune, 32*(4), 117–126. [In Chinese].

Jin, H., & Cai, W. (2022). Understanding the smartphone usage of Chinese outbound tourists in their shopping practices. *Current Issues in Tourism, 25*(18), 2955–2968.

Jin, X., & Sparks, B. (2017). Barriers to offering special interest tour products to the Chinese outbound group market. *Tourism Management, 59*, 2015–2215.

Jin, X., Wu, L., Becken, S., & Ding, P. (2016). How do worry, self-efficacy and coping interact? Examining Chinese tourists to Australia. *Journal of China Tourism Research, 12*(3–4), 374–393.

JLL. (2019, January 18). *Australian luxury hotels up their game for Chinese tourists*. Accessed April 22, 2022, from https://www.jll.com.co/en/trends-and-insights/cities/australian-luxury-hotels-game-chinese-tourists

Kaplan, S. (2023, January 27). WeChat tightens Beijing's grip on Chinese diaspora. *Wall Street Journal*.

Kashiwagi, S., Nagai, H., & Furutani, T. (2018). VFR travel generated by international students: The case of Japanese students in Australia. *Tourism: An International Interdisciplinary Journal, 66*(1), 89–103.

Keane, M. (2017). Foreword. In M. Kent, K. Ellis, & J. Xu (Eds.), *Chinese social media. social, cultural, and political implications* (pp. 1–3). Routledge.

Kim, J., & Fesenmaier, D. (2017). Sharing tourism experiences: The post trip experience. *Journal of Travel Research, 56*(1), 28–40.

King, R., Racherla, P., & Bush, V. (2014). What we know and what we don't know about online Word-of-Mouth: A review and synthesis of the literature. *Journal of Interactive Marketing, 28*, 167–183.

Kohler, A. (2017, October 28). Daigou shoppers ring up the sales. *The Australian*.

Koob, S. (2017, October 28). China girls drive the Aussie tourism boom. *The Australian*.

LEK. (2017). *The China tourism economy: Reaching Australia's potential.* Australia China Business Council.

Li, F. (2020). Understanding Chinese tourists' motivations of sharing travel photos in WeChat. *Tourism Management Perspectives, 33*, 100584.

Lin, Y., Lee, Y., & Wang, S. (2012) Analysis of motivation, travel risk, and travel satisfaction of Taiwan undergraduates on work and travel overseas programmes: Developing measurement scales, Tourism Management Perspectives, 1(2), 35–46.

Litvin, S., Goldsmith, R., & Pan, B. (2008). Electronic word-of-mouth in Hospitality and Tourism Management. *Tourism Management, 29*(3), 458–468.

Liu, C. (2021). Exploring selfie practices and their geographies in the digital society. *Geographical Journal, 187*, 240–252.

Liu, Y., Bao, J., & Chen, K. (2017). Sentimental features of Chinese outbound tourists in Australia: Big-data based content analysis. *Tourism Tribune, 32*, 46–58. [In Chinese].

Liu, X., Mehraliyev, F., Liu, C., & Schuckert, M. (2020). The roles of social media in tourists' choices of travel components. *Tourist Studies, 20*(1), 27–48.

Ma, Y., Hardy, A., & Ooi, C. (2020). Researching Chinese Tourists on the Move. *Journal of China Tourism Research, 16*(2), 214–229.

Magno, F., & Cassa, F. (2018). The impact of social media influencers in tourism. *Anatolia, 29*(2), 288–290.

McCartney, G., & Pek, R. (2018). An examination of Sina Weibo travel blogs' influence on sentiment towards hotel accommodation in Macao. *Journal of China Tourism Research, 14*(2), 146–157.

McHugh, B. (2017, June 15). Australia must be China-ready to capitalise on its unique tourism. *ABC Rural.*

Miao, Y. (2015). The influence of Electronic-Wom on tourists' behavioral intention to choose a destination: A case of Chinese tourists visiting Thailand. *AU-GSB e-JOURNAL, 8*(1), 13–31.

Munar, A., & Jacobsen, J. (2014). Motivations for sharing tourism experiences through social media. *Tourism Management, 43*, 46–54.

Murphy, D. (2017, May 13). Middle-income Chinese are the new wave of tourism. *Sydney Morning Herald.*

Nusair, K., Bilgihan, A., Okumus, F., & Cobanoglu, C. (2013). Generation Y travelers' commitment to online social network websites. *Tourism Management, 35*, 13–22.

Oliveira, T., Araujo, A., & Tam, C. (2020). Why do people share their travel experiences on social media? *Tourism Management, 78*, 1–14.

Pearce, P., Wu, M., & Chen, T. (2015). The spectacular and the mundane: Chinese tourists' online representations of an iconic landscape journey. *Journal of Destination Marketing and Management, 4*, 24–35.

Peng, Y. (2019). Sharing food photographs on social media: Performative Xiaozi lifestyle in young, middle-class urbanites' WeChat 'Moments'. *Social Identities, 25*(2), 269–287.

Pourfakhimi, S., Duncan, T., & Coetzee, W. (2020). Electronic word of mouth in tourism and hospitality consumer behaviour: State of the art. *Tourism Review, 75*(4), 637–661.

Prayag, G., Hall, C., & Wood, H. (2018). I feel good! Perceptions and emotional responses of Bed and Breakfast providers in New Zealand toward Trip Advisor. *Journal of Hospitality Marketing and Management, 27*(1), 1–20.

Qi, S., & Chen, N. (2019). Understanding Macao's destination image through User-generated content. *Journal of China Tourism Research, 15*(4), 503–519.

Sann, R., Lai, P., & Chen, C. (2021). Review papers on eWOM: Prospects for hospitality industry. *Anatolia, 32*(2), 177–206.

Sier, J. (2017, October 6). Daigou retailer sells well on ASX debut. *Sydney Morning Herald.*

Skavronskaya, L., Moyle, B., & Scott, N. (2019). Collecting memorable tourism experiences: How do 'wechat? *Journal of China Tourism Research, 16*(3), 424–446.

Smith, M. (2016, December 28). Delivering the real experience. *Australian Financial Review*, 42–43.

Song, S. (2022, June 29). RED (Xiaohongshu) trends 2022: emerging trends in fitness, fashion & beauty. *China Digital Times.* https://walkthechat.com/red-xiaohongshu-consumer-trends-2022

Sotiriadis, M., & van Zyl, C. (2013). Electronic word-of-mouth and online reviews in tourism services: The use of twitter by tourists. *Electronic Commerce Research, 13*, 103–124.

Su, M., Wall, G., Wu, B., et al. (2021). Tourism place making through the bioluminescent 'Blue Tears' of Pingtan Islands, China. *Marine Policy, 133*, 104744.

Sun, M., Ryan, C., & Pan, S. (2015). Using Chinese travel blogs to examine perceived destination image: The case of New Zealand. *Journal of Travel Research, 54*(4), 543–555.

Tan, S. (2017, October 6). Golden week for tourism, shops. *Australian Financial Review.*

Tham, A., Zhang, X., Spinks, W., & Wang, L. (2022). Chinese outbound tourists and *Daigou* as soft power. *Journal of China Tourism Research, 18*(6), 1143–1163.

Tourism Western Australia. (2011). *Our direction in China 2012–2015.* Tourism Western Australia.

Urbain, J.-D. (1989). The tourist adventure and his images. *Annals of Tourism Research, 16*(1), 106–118.

Victoria The Education State. (2018). *Engaging with Chinese tourists.* Victoria State Government.

Walsh, M., Dale, N., & Johns, R. (2019, December 27). #travelgram: Live tourist snaps have turned solo adventures into social occasions. *The Conversation.*

Wang, Z., & Connell, J. (2021). Taiwanese working holiday makers in rural and regional Australia: Temporary transnational identities and employment challenges. *Australian Geographer, 52*(2), 191–207.

Wang, D., Park, S., & Fesenmaier, D. (2016). The role of smartphones in mediating the touristic experience. *Journal of Travel Research, 51*(4), 371–387.

Wei, Y., & Wu, T. (2021). Visual representation of a linear tourist destination based on social network photos: A comparative analysis of cross-cultural perspectives. *Journal of Tourism and Cultural Change, 19*(6), 781–804.

Wei, Y., Zhang, G., & Weng, W. (2018). The influence of 'Internet Plus' on China's tourism: Evolution and breakthrough. In *Proceedings of 4th international conference on behavioral, economic, and socio-cultural computing (BESC) July 1, 2017, BESC* (pp. 1–5).

Weiler, B., & Chen, H. (2016). Repositioning Australian cities as settings for memorable nature-based experiences aimed at Chinese tourists. *Tourism Recreation Research, 41*(3), 246–258.

Weiler, B., & Yu, X. (2006). *Understanding experiences of Chinese visitors to Victoria*. Sustainable Tourism Cooperative Research Centre.

Wilk, V. (2019, June 21). How a humble Perth boathouse became Australia's most unlikely tourist attraction. *The Conversation*.

Wong, J., Lai, I., & Tao, Z. (2020). Sharing memorable tourism experiences on mobile social media and how it influences further travel decisions. *Current Issues in Tourism, 23*(14), 1773–1787.

Wu, M. (2015). Driving an unfamiliar vehicle in an unfamiliar country: Exploring Chinese recreational vehicle tourists' safety concern and coping techniques in Australia. *Journal of Travel Research, 54*(6), 801–813.

Wu, M., & Pearce, P. (2014). Chinese recreational vehicle users in Australia: A netnographic study of tourist motivation. *Tourism Management, 43*, 22–35.

Wu, M., & Pearce, P. L. (2016). Tourism blogging motivations: Why do Chinese tourists create little 'Lonely Planets'? *Journal of Travel Research, 55*(4), 537–549.

Xia, J., Adriano, R., Zea, M., Crowe-Delaney, L., et al. (2018). *Are we China ready? Chinese tourism in Western Australia*. BCEC Research Report No. 14/18.

Xiang, Z., & Gretzel, U. (2010). Role of social media in online travel information search. *Tourism Management, 31*, 179–188.

Xu, H., Ding, P., & Packer, J. (2008). Tourism research in China: Understanding the unique cultural contexts and complexities. *Current Issues in Tourism, 11*(6), 473–491.

Yoo, K., & Gretzel, U. (2011). Influence of personality on travel-related consumer-generated media creation. *Computers in Human Behavior, 27*(2), 609–621.

Yu, H., Xu, J., & Sun, P. (2022). Introduction: Platformization of entrepreneurial labor via Chinese digital networks. *Global Media and China, 7*(3), 253–262.

Zarezadeh, Z., Rastegar, H., & Gretzel, U. (2018). Reviewing the past to inform the future: A literature review of social media in tourism. *Czech Journal of Tourism, 7*(2), 115–131.

Zeng, B., & Gerritsen, R. (2014). What do we know about social media in tourism? A review. *Tourism Management Perspectives, 10*, 27–36.

Zhang, Y., & Peng, Y. (2014). Understanding travel motivations of Chinese tourists visiting Cairns. *Journal of Hospitality and Tourism Management, 21*, 44–53.

CHAPTER 5

Campus Tourism Meets Harry Potter

This chapter extends beyond the previous chapter through examining how the rapid rise of social media in the 2010s created a more individualistic tourism. Instantaneous electronic word-of-mouth (eWOM) enabled tourists to choose their destinations based on the recommendations of others: a phase increasingly marked by the rise of bloggers and influencers and the search for 'authenticity'. This more individualistic tourism is then elaborated through a focus on two interlocking niches: education or campus tourism and the particular and highly unusual version of that which drew tourists to the University of Sydney in search of traces of Harry Potter. We therefore initially focus on campus tourism—something of a Chinese phenomenon—introduce popular culture and tourism and then narrow these strands down to the Sydney University context, and the eventual emergence of Hogwarts tourism.

Over time, in the context of tourism at the University of Sydney, individual tourists drew attention to particularly enjoyable and photogenic 'marked places' (打卡点: *daka* destinations and sites) that should be visited, and which many claimed to have discovered or seen in a new light. Chinese tourists have an extremely high level of social media use and a distinctive commitment to, and interest in, sharing tourism experiences (Wong et al., 2020) hence, as in China (see Chap. 6), *daka* sites also emerged in Australia. We illustrate this through an examination of the social media practices of Chinese tourists, before, during and after visits to

J. Connell et al., *Chinese Tourism in Australia*,
https://doi.org/10.1007/978-981-97-2477-2_5

Australia. No prior studies have analysed social media practices that, as here, suffuse sites with heritage value despite having no 'genuine' basis for this. At the same time therefore, we evaluate tourism at one such place—the first *daka* site in Australia—through what we have called soft heritage: fictitious sites of popular cultural tourism, unrelated to 'authentic' sites of production, that emerge through culturally intermediated social media practices, which have occasionally facilitated vast distributed networks of connected tourists and 'marked places'. Other similar sites include the Australian Standing Stones at Glen Innes (NSW) where, in this case deliberately, the site was constructed in the 1990s to enhance the region's perceived Celtic heritage and stimulate tourism; tourists at the site have discerned elements of that heritage and authenticity despite its recency (Connell & Rugendyke, 2010). Broadly similar, and also deliberately constructed, replica sites elsewhere have been described as 'staged authenticity' or 'copysites' (Yousaf & Fan, 2020). This chapter traces and accounts for the rise of one unusually distinctive 'marked place'—the 'Harry Potter building' at the University of Sydney, previously absent from conventional tourism in Sydney and hitherto unknown to residents of Sydney. The long-term popularity of this site eventually led to its being discussed in conventional city guidebooks (such as Lonely Planet). We examine why social media were successful in directing tourists, why this particular site was attractive and how it became a key component in the creation of a new Chinese tourist geography of Sydney and of Australia (Chap. 6).

Campus Tourism

In this century a distinctive campus tourism has emerged. Campus tourism has thrived in China and international tourism from China has fostered, or—as in the United States—contributed to, the growth of this form of niche tourism elsewhere. Oxford and Cambridge, the Ivy League universities of the United States and the mediaeval universities of Europe are all recipients and more arguably beneficiaries of campus tourism. Campus tourism may even be traced to 'the seventeenth century fashion, the "Grand Tour", undertaken by young aristocrats' (Connell, 2000, p. 2), with the phenomenon of tourists visiting institutions of higher learning mainly involving older universities with an architectural heritage (Albino, 2015; Woodward, 2013; Mangi et al., 2019; Hayward, 2020). In Australia, the University of Sydney attracts many visitors, both organised tour groups and independent tourists, with the majority coming from

China. For tourists, a distinctive tourist experience combines physical and cultural factors: heritage architecture, mythology, open spaces and educational status, experienced in multiple ways and combinations. Social media have shaped both the diversity of visiting and the perceptions of university sites, an incidental and experiential complexity that is both of benefit and a challenge to campus tourism. In many cases, exemplified by Sydney University, this niche tourism experience has evolved with neither marketing nor deliberate creation and curation of sites and activities, despite its potential dual benefits to the University: immediate revenue and future international students. Tourists largely create both context and experience, incidentally challenging the otherwise seemingly ubiquitous notion of co-creation, where tourists and host institutions supposedly interact to jointly create a tourist place and experience (McManus et al., 2021). Here campus tourism was only co-created by default.

Campus tourism constitutes visiting by organised tour groups or individuals to a university primarily with an aesthetic and/or educational objective, unrelated to accommodation, employment, business or immediate education. People, often parents of potential university students, become campus tourists in response to varying combinations of reputation, prestige, heritage architecture and landscaping, partly related to both the tangible and intangible cultural assets of a university. That has emerged from an earlier era when 'campus-based tourism [wa]s beginning to develop as a significant market sector in tourism' (Connell, 1996, p. 541) with financially strapped universities utilising their accommodation resources during long vacations. Campus-based tourism then focused on income generation from underused assets and was not necessarily related to any physical or educational characteristics of universities. Universities as tourist destinations within a broader cultural and urban tourism were incidental. Despite predictions of a twenty-first century growth in university campus tourism, in practice university marketing of accommodation proved of no great interest (other than for conferences) and consequently, outside China, 'campus tourism has generally been the subject of little academic interest' (Woodward, 2013, p. 265). No distinctive category was then in existence and numbers were few and inconsequential.

A distinctive campus tourism emerged in the twenty-first century, although visiting institutions of higher learning by tourists was scarcely new. Historic buildings, elegant landscapes, a reputation for educational excellence over many centuries and even an intellectual aura were common hallmarks of the most visited universities, such as Oxford and

Cambridge, Harvard, Salamanca in Spain, Heidelberg in Germany and the Al-Azhar University and Mosque in Cairo (Woodward, 2013). In the late twentieth century universities in many countries began to offer tours for prospective students, and their families; they were effectively marketing the university courses, residences and the wider environment, but the tours were not designed for a general public. A dichotomy had thus emerged between visiting universities as places of cultural heritage (much like, for example, visiting cathedrals) and visiting universities as potential sites of future education; the former did not need to be promoted and only concerned a few historic universities, with both domestic and international visitors, while the latter was bound up with marketing but also primarily concerned relatively few prestigious universities, and mainly involved potential domestic students. That changed as international education became more desirable and sought after.

In the twenty-first century, particularly through Chinese visitors who initially travelled in organised tour groups and were visibly distinct from other visitors, staff and students on campus, the phenomenon of university campus tourism acquired a new international flavour. The eventual international trend followed the emergence of university campus tourism in China, which allegedly began in 1996 when Haidian District Tourism in Beijing organised a 'Pursuit of Knowledge' tour to Tsinghua University for 100 distinguished middle-school students. Visiting campuses subsequently became an even more popular activity and campus tourism an important component of domestic tourism (Zhou, 2009). Two decades later university campus tourism, especially in summer, was said to have 'long been a tradition for many Chinese. Large numbers of tourists visit university campuses—some of them hundreds of years old—either for their scenic beauty or the historic and culturally rich buildings' (Chenyu, 2018). A general rise in family incomes and increasing attention to, and enthusiasm for, the education of younger generations supported the growth of campus tourism (Zhou, 2009; Ling & Dai, 2012). By the summer of 2007, Tsinghua University and Peking University were said to average 4000 tourists every day, with peaks reaching 10,000 tourists (Li, 2013). In the summer of 2017, a record 430,000 Chinese and foreign tourists visited the two universities (Zhang, 2017). Such substantial numbers meant that 'campus tourism [became] a controversial issue and profoundly debated in China' (Mangi et al., 2019, p. 4). As an emerging and distinctive version of over-tourism, campus tourism posed problems. On a much smaller scale, as campus tourism emerged and evolved in Australia for much the same reasons, similar problems were eventually encountered.

Chinese university tourism took on a more educational flavour as the middle class grew, higher education became in much greater demand, students became more aspirational and they and their families sought to visit potential destinations to pursue future educational goals. Prominent Chinese universities had little need to advertise or market themselves and were sometimes overwhelmed by the enormous number of visitors. Indeed, a challenge for Chinese university campus tourism was that tourists did not always follow recommended 'rules', affecting teaching, so increasing potential safety risks on campus. In 'Peking University, teachers and students complained about noisy crowds disrupting the academic atmosphere'. That proved a key tension because 'people come to such campuses aiming to experience cultural education heritage, interact with academic communities, and give educational awareness to their children' (Mangi et al., 2019, pp. 4, 21). The most prestigious campuses experienced distinctive environmental pressure, especially on facilities such as campus cafeterias, so that 'increasing public access to university campuses by tourists, including families of prospective tourists, are creating such levels of disturbance that the sense of place and atmosphere desired by students and faculty is being lost' (Woodward, 2013, p. 270). Some tour guides (who were not university personnel) were poorly trained and could not provide an accurate and valuable experience. Ultimately universities were concerned to avoid a situation where they appeared to be inhabited by tourists rather than scholars. Regulation and management were required, and several universities capped numbers on weekdays (Chenyu, 2018). Nonetheless, because of the rise of a new middle class with educational aspirations, leisure time and money, Tsinghua University and Peking University became the preferred destinations of campus tourists.

As campus tourism grew in popularity in China it was exported as a practice by the aspirational middle class; it grew from the interests and enthusiasm of tourists, keen to secure educational opportunities for family members, rather than from active promotion by universities (Zhang, 2017; Chenyu, 2018; Yuming, 2018; Mangi et al., 2019). Universities that ranked highly in international schemes, replete with heritage architecture and close to the city centres, benefited most. Chinese interest in Chinese universities quickly spilled over into international destinations, in North America and Europe; it was suggested that for Yale 'international prestige makes it a must-see landmark for many Chinese tourists' (Westfall,

2017). Chinese tourists in Thailand and Korea also sought out what they perceived to be the most prestigious universities, and which accorded them status (O'Regan & Choe, 2022). While group package tours included universities that were sites of heritage and culture, most of the growth in numbers of contemporary campus tourism came from Chinese families visiting universities, that could be places for the future education of family members. It was not a phenomenon that spilled over into other ethnic and national groups, despite the global rise of international education.

Western universities were as unprepared for this new international campus tourism as their Chinese counterparts. They were reluctant to market themselves and engaged in management primarily as a means of regulation and directing numbers, if tourism impinged on other values. Indeed, campus tourism in China and internationally is conspicuous in being a rare form of niche tourism that has grown almost entirely without marketing and, in order to avoid over-tourism, management has tended to constrain, channel and reduce numbers rather than seek greater numbers. Universities preferred to restrict opening times to visitors, rather than develop effective marketing strategies; many consciously chose not to do so. Contemporary international campus tourism emerged from tour companies, and the Chinese tourists themselves, rather than the universities. Extensive use of social media conveyed tourist experiences, particularly through the instantaneous posting of images (see below). Efforts to capture short-term financial benefits by universities were usually foregone in favour of the longer-term circulation of images and awareness which was anticipated to raise international student enrolments. Nonetheless, seeking more international students, many universities eventually engaged in what amounted to subtle forms of place marketing, as long as they did not disrupt the peaceful ambience. However, conflicts between teaching and tourism were ubiquitous (O'Regan & Choe, 2022).

International university campus tourism is linked to ease of travel, belated promotion by universities, geopolitical concerns and the shifting interests of future scholars and their relatives. Managing tourist numbers and impacts while maintaining a focus on education was a constant challenge as university campus tourism expanded: a rare niche where promotion and marketing are almost exclusively undertaken by the tourists, representing a distinctive form of tourism largely initiated, led and shaped by the tourists themselves. We now examine the unusual circumstances of campus tourism at the University of Sydney and its interactions with other tourism activities.

The rapid rise in Chinese outbound tourism contributed to the popularity of campus visits by Chinese tourists, notably at the University of Sydney, since Sydney was the dominant destination in the early years of Chinese tourism and was the best-known university in China. As Chinese tourism evolved from the dominance of group tours to more independent travel, a distinctive feature became the increased use of internet platforms to direct tourism. Their role before, during and after in the creation of 'marked places' to which Chinese tourists should go, and even provide photographic evidence of their visit, gave campus tourism in Australia a distinctive flavour. Such circumstances were not however related only to campus tourism. Tourism at the University, although displaying many of the components of campus tourism elsewhere, has taken an unusual direction and character centred on the 'soft heritage' of Hogwarts and Harry Potter.

Popular Culture … in a University

What began as campus tourism acquired a more distinctive and, in an Australian context, a unique form of popular culture tourism. Here campus tourism has intersected with another dimension of tourism studies related to both popular culture and heritage, where tourists infuse the university with heritage values associated with popular culture, but where there is no factual basis for them.

Popular cultural tourism is well established for global cultural 'brands' (such as *Game of Thrones*, Elvis Presley and Harry Potter), and the desire to visit key production sites, take photographs and recreate scenes from films or television episodes for personal posts can fuel substantial increases in tourist numbers outside formal marketing campaigns (Lexhagen et al., 2013; McWha & Beeton, 2018). More recently social media has generated new geographies of tourism and mobility entirely unrelated to the 'authentic' sites of popular culture production: when hitherto unknown sites emerge as tourist destinations through practices of social media use focusing on a place's physical characteristics and endowing it with a distinct but invented heritage value, linked to fictional scenes or characters from popular culture.

All forms of popular culture including music, television, film, art and literature have inspired tourism and destination marketing (e.g. Beeton,

2016; Gibson & Connell, 2005). While tourism arising from popular culture takes many forms, including immersive and active participation (such as attending live theatre or music performances), among the most pervasive are visits to 'authentic' historic sites of birth, life and death (including Graceland, or Jim Morrison's grave in Père-Lachaise cemetery, Paris), scenes depicted in paintings (including street art and rock art), books, films and television shows (like the *Bridges of Madison County*, the Edinburgh of John Rebus and the 'home' of Anne of Green Gables in Prince Edward Island), or multiple anime sites (e.g. Tung et al., 2017), and the Northern Irish and Croatian sites of *Game of Thrones* (e.g. Mannheimer et al., 2022; Waysdorf & Reijnders, 2018a) and places of production (such as film studios or where famous novels were written, like the Brontë sisters' home in Haworth, Yorkshire) (e.g. Gothie, 2016; Zhang & Ryan, 2018). Various strategies among tourism promoters and media and entertainment corporations seek to formalise such sites as 'authentic' heritage destinations, from the museumification of homes to the conversion of film studios into theme parks. The latter include Warner Brothers' studio complex at Leavesden in England, where the Harry Potter films were made, and 'Hobbiton' in New Zealand, where *The Lord of the Rings* was filmed (Le Heron, 2020). Tourists are drawn to visit and see the 'real' scenery, residential homes, production facilities, stage sets and costumes.

However, in some contexts and among specific cultural groups, heritage values become associated with sites of social media users, where there is no such claim to authenticity, such as Whitby in Yorkshire being the supposed holiday resort of Bram Stoker's Dracula (Spracklen & Spracklen, 2014). Sydney University is such a case. A university otherwise absent from tourism circuits, in a city with multiple attractions (the Sydney Opera House and Harbour Bridge, Bondi and other world-renowned beaches) has become a marked place among Chinese tourists through an invented association between one of its buildings and the Harry Potter books and films. That association was concocted by neither tourism marketers nor university managers. Rather, social media users themselves generated the association, by conflating architectural heritage (a graceful old building) with popular cultural meanings. From what one Chinese tourist, quoted below, described as a 'beautiful misunderstanding', a popular tourist destination emerged. This, we contend, amounts to a hitherto under-acknowledged form of soft heritage: the generation of tourist sites that, notwithstanding absent claims to authenticity, fuel cultural meanings and elicit emotional responses.

We examine how, not quite by chance, but without internal interest or promotion, the University of Sydney emerged as a distinct destination, and one of the city's most prominent 'marked places', specific to Chinese visitors but 'unknown' to others. Reasons for visiting that we document empirically below include group tours, education, heritage and photography. Even more important however, and centred around each of these, is the 'Harry Potter building': a Victorian-era neo-Gothic structure on campus that was not involved in the making of the films, nor previously an established tourist destination, but which bore a more than passing resemblance to the Hogwarts School of Witchcraft and Wizardry from the film adaptations of JK Rowling's best-selling books. Occasionally that extended into stories of Harry Potter being a former student, or JK Rowling being a medical school graduate there, or the various Potter films being made there. The 'soft heritage' built around the myth of Harry Potter being filmed at the campus, positioned the University of Sydney as a marked place for many Chinese tourists with one corner of the Quad becoming the 'Harry Potter Building' (a nomenclature long familiar to Chinese students on campus but unknown to most others on campus). Embellishments, distortions and supposed links to actual local history occur at other similar sites, such as New Zealand's Hobbiton, but are obviously impossible since Harry Potter is a fictional character (if anything, with British roots, where a more 'authentic' film tourism exists (Lee, 2012)). In a city of multiple global attractions, social media use combined with architecture, ambience and Chinese enthusiasm for Harry Potter in creating a new, culturally-inflected tourism geography: an emergent soft heritage.

Harry Potter in China

JK Rowling's best-selling Harry Potter books and especially the films are well-known and much enjoyed in China, with both Chinese language and English copies being read. The first Chinese version was produced in 2000 and the books topped children's book sales in China in almost every subsequent year. Some entered school curricula. All eight films have run there. Twenty million copies of the books were sold in China before 2017, largely paralleling the rapid growth and consumption of a middle class, to whom the books most appeal. That has sparked a 'Pottermania', attracted by a parallel fantasy world, lacking in the more formal Chinese literature, of creativity, escapism and individualism where a sense of autonomy and empowerment stimulated some to rethink their own lives (Erni, 2008,

2013). The seven Potter novels took on somewhat greater international themes over time, but the presence in four of the films of an actress of Chinese ethnicity, Katie Leung, who played Cho Chang, a witch who attended Hogwarts School and was Harry Potter's first love interest, further boosted their attraction. That translated into a significant demand to see the Harry Potter building, as the books and films structured everyday touristic practices, and diversified interlocking objectives.

Tourism at the University of Sydney?

Like most universities, Sydney University has never sought to indulge in, let alone market, campus tourism. The University, the oldest university in Australia, with a historic core building, the Quadrangle, initially constructed in the 1850s, is nevertheless replete with architectural heritage. The sandstone structure, designed by the former NSW Colonial Architect, and subsequently University architect, Edmund Blacket, replicates Oxbridge colleges. The site is now an inner-city location within Australia's largest city, yet the University has been almost completely ignored in standard guidebooks. The Lonely Planet guide to New South Wales had a short paragraph on the Nicholson Museum ('one of the city's great under the radar attractions') but made no mention of the University of Sydney, despite the museum then being in the Quad (Lonely Planet, 2017, p. 97). Foreign-language guidebooks on Australia, being less comprehensive, never mentioned the University.

The Insight Guide referred to the University as 'a leafy haven built in self-conscious Oxbridge style' (2016, p. 114) but did so in a paragraph that discussed travelling between the adjoining suburbs of Glebe and Newtown rather than as a distinct place to be visited. Both Eyewitness Australia (2018) and Fodor's Essential Australia (2019) made no mention of the University. The Rough Guide (2017, p. 133) however remarked that 'a path and steps lead up into Sydney University, inaugurated in 1850, your gaze is led from the walkway up to the Main Quadrangle and its very Oxford-reminiscent clock tower and Great Hall. You can wander freely around the university grounds, and there are several free museums and galleries to visit'. Collectively, such brief references to the University were primarily a function of competing metropolitan attractions. In 2018 there was, however, a distinctive change; the most recent Lonely Planet guide to Sydney stated:

> Flanked by three grand halls that wouldn't be out of place in Harry Potter's beloved Hogwarts, the Sandstone Quadrangle has a Gothic Revival design that dips its mortarboard to the stately colleges of Oxford (Symington, 2018, p. 157).

What had contributed to the new and different coverage of the university was the presence of Chinese tourists 'discovering' the Quad, infusing it with heritage values associated with Harry Potter, social media establishing its destination status and a wider Australian public becoming aware of the connection.

Since the mid-2010s the University had become a destination for many Chinese tourists, with group package tours taking place at least daily, and itineraries usually booked in China through Ctrip or another agency (which usually included the standard iconic sites). Chinese tourists easily constituted the largest source of international tourists at the University, visibly as groups, making up several hundred a day. No other Sydney tours included the University and Chinese tourists were the only group taking package tours there. Many of those who took the package tours did not necessarily therefore have a particular objective of visiting the University, which was simply a stop between the city centre and the Blue Mountains. For some it was undeniably a toilet stop. Beyond that were growing proportions of independent travellers, also from China, usually making up smaller numbers. Peak seasons correlated with holidays in China, especially the December–January winter vacation, and the June–August summer vacation. Both Chinese tour groups and individual travellers visited the University without charge (with a University assumption that long-term benefits would accrue through an eventual increase in the number of Chinese international students). Although the University engaged in no tourism marketing, visitor numbers steadily increased, assisted by the University having more overseas Chinese students than any other Australian university (with 70% of all overseas students coming from China) so that it was increasingly better known there. By 2015 there were almost 6000 Chinese students at the university.

Like other contemporary universities the University of Sydney was not averse to marketing, but marketing was determinedly academic. The University of Sydney website has offered 'heritage tours', a one-hour tour that focuses on the Quad and nearby buildings, and a self-guided tour with a downloadable PDF guide (in six languages) which centres on the Quad but covers a greater range of buildings. The self-guided tour was focused on the diversity of the university and featured recreational facilities and residential colleges. The website proclaimed: 'Whether you're a tourist, history buff, prospective student or simply curious to learn more about our campus, we run a range of tours to cater for all' but, as elsewhere, tours were aimed at prospective students and families who wished to visit the university in their own time, and at their own pace. Participants were few. University tourism was effectively driven by the tourists themselves, who contented themselves with wandering around without formal direction, and by social media.

At the same time, independent tourists increasingly made more individual decisions to make the University a destination, as Chinese social media promoted it as a destination. A single post on the influential Chinese travel and tourism platform, *Mafengwo*, provided some information about the history, status, location, architecture and recreational opportunities for visitors to the University of Sydney:

> University of Sydney is the first university in Australia, and is one of the best universities around the world. ... Many buildings in the campus retain the traditional architectural style, representing the cultural features of Australia. The campus is surrounded by parks and a lake, and is open to the public. It is a peaceful place in which to relax (*Mafengwo*, 10 October 2018)

This website also included almost 11,000 photographs of the University of Sydney, mostly of the Quad and nearby sites, that highlighted historic architecture, blue sky, lawns and colourful jacaranda trees in bloom. Chinese tourists have responded to posts on this platform and on other social media, with posts on the social media site, Little Red Book, having become the main source of information about tourism at the University, through recommending that the university be included in tourist visits Sydney and by stressing what to see and do on the campus. Quite unintentionally, the University became a popular and distinct tourist site with visitor numbers growing through the 2010s. To assess the significance of this, its impact and the perceptions of the tourists, we began a study focused around the Chinese tourists at the University.

Talking with Tourists

At the core of our study of campus tourism at the University of Sydney was seeking the experiences of tourists at the university, while also reviewing the perspectives of tourism providers. Undertaking cross-cultural research proved challenging, amplified by the time constraints of tour groups, as they met tight schedules, language issues and political sensitivities. The mixed-methods approach employed involved on-site structured interviews with Chinese tourists in and around the Quad, in-depth key informant semi-structured interviews, participant observation (including a Chinese language bus tour of Sydney's tourist attractions), and tagging on to tours at the University, alongside the analysis of Chinese language social media. That later extended to visiting various marked places in and around Sydney, such as Bare Island and Diamond Bay (see Chap. 6). Mixing methods enabled us to obtain information and perspectives from tourists and university management, to engage directly with tourists, to position this engagement within a longer timeframe through the analysis of social media and to triangulate findings and strengthen the rigour of the research. All this was supplemented by informal observations at the University over a number of years. Indeed, it was those informal observations that initiated our research programme.

This multi-faceted approach reduced some of the complications and challenges of interviewing tourists on tight schedules and ensured that quantitative data was obtained, while encouraging in-depth qualitative research when possible. The interviews were undertaken by the female author who was a native Mandarin speaker and knowledgeable of both Chinese cultures and the university context. Nonetheless tourists were often concerned and puzzled at being the subject of interviews and time was limited to ease concerns. However while package tourists were hurried, more independent tourists had more time and interest in being involved.

Between April and July 2019, 74 interviews were conducted with Chinese visitors to the University of Sydney, including three interviews with tour guides. Some interviews included multiple participants, thus we interviewed 85 tourists (62 female and 23 male), with the length of interviews varying based on the time-sensitivity of the participants. The sample included three participants from Taiwan, and one from Hong Kong, with the vast majority coming from various cities in mainland China. Observation of tourism over much longer periods suggested that the sample was representative both of tourists to the University and to Sydney as a whole. We concluded the formal survey when we were confident that we had reached saturation with regard to important new information being provided by interviewees.

In addition to Chinese tourists, we interviewed University of Sydney managers who had some relationship to tourism provision, to understand the role of the university administration and discuss issues relating to the evolution of campus tourism. The interviews were audio recorded when consent was granted and later relevant sections were transcribed. This was supplemented by analysis of both Chinese and English language social media, by analysing tourist guidebooks (in English and Chinese), all couched within informal observations over many years.

These interviews were followed by our social media approach focusing on the apps used by Chinese tourists, including Weibo, *Mafengwo, qiongyou*, and video sharing platforms such as bilibili, vlog and Little Red Book (*xiaohongshu*). Initially, posts over a two-year period (2018–2019) were collected and counted to understand the extent of posting patterns. They were then analysed for their content by identifying the key words (translated if necessary) and themes present. Further analysis included counting the number of likes various posts received and contextualising the date of the post with other events that could influence the nature of the post. The posts were also cross-referenced with the more qualitative of the on-site interviews to develop and integrate key themes from both research techniques. Online posts were not just useful adjuncts to the interviews but proved vital in establishing more detailed and subtle perspectives on motivations for travel, visiting the University, the role of photography (and the value of smartphones) and reflecting on and recounting the experience. It became readily evident that suggestions made during travel could result in tourists diverting from their intended routes and activities to initiate new and unplanned activities (Bronner & de Hoog, 2011; Hwang, 2011) and themselves updating friends, relatives and many others on their own ongoing tours. This had particular significance for the evolution of Chinese tourism in Sydney, as it extended beyond campus tourism (Chap. 6).

Within this sample of tourists we identified four loose clusters. There was inevitably some overlap between these clusters, related to how aspirations varied—indeed sometimes on the spot—and were often multiple and diverse, sometimes according to where being at the University fitted into a wider schedule or tourism structure. A third of the tourists came with a group tour on which the University of Sydney was a scheduled destination; visiting the University may not therefore have been a reason for joining the tour. Of the remaining tourists, all of whom specifically intended to visit the University, some 21% (12/57) of them came with primarily educational objectives, almost 25% (14/57) of tourists were primarily VFR tourists, visiting the University because they had relatives studying there (or potentially studying there) and 54% (31/57) of tourists came to see the main Quad—and especially the Harry Potter building. The present study focused on these independent tourists because of their ability to make independent choices of destination and be flexible.

The three groups of independent tourists overlapped considerably, having mixed motives that linked education, heritage buildings and architecture and the Harry Potter building. All tourists expected to take photographs on the site. Campus tourism at Sydney University thus reflected a range of often interrelated motives and possibilities but was centred on people interested in future education opportunities for themselves or other family members and/or interested in heritage, whether historic buildings or the myth of Harry Potter being filmed there or because the buildings appeared similar to Hogwarts.

Skewing of this structure of visitation was related to age. Of the tourists coming to see the Harry Potter building, only 9% (3/31) were over 30 years of age, reflecting the popularity of Harry Potter books and films among younger people. By contrast, in the cluster of interviewees who were on a scheduled group tour, 71% (20/28) of participants were aged 50 years of age, a partial outcome of the challenges of international travel for older tourists. Only two Chinese tourists had previously not travelled overseas, emphasising that Australia was generally a destination for Chinese tourists after they had visited other parts of the world. We can now consider the perspectives of these tourists in greater detail.

Chinese Tourists at the University of Sydney

The most distinctive and very visible group of tourists were those on group tours where the University was one stop of many on a day or half-day tour. Such package tours were analogous to similar tours in China. Packages from various locally based Chinese travel companies included the University in a network of sites south of the harbour. Some group tourists were not necessarily interested in the University, but it was part of a tour that included various well-known inner-city sites, extending to the Blue

Fig. 5.1 Group tour at University of Sydney Quad

Mountains. Tourists sometimes had little more than 15 minutes at any site, which meant a limited time for observation, absorbing information about the University, reflection or photography. Limited time restricted the potential for exploration beyond the Quad, but still afforded prime photo opportunities (see Fig. 5.1). Guides (usually relatively recently arrived local residents of Chinese origin) were often not well-informed and information was basic and many of those on group tours had little to say about the experience. Group tourists were often poorly informed about the University but they had no great expectations. Independent tourists were much better informed.

A second group were those who were interested in the university as a heritage site (of elegant older buildings) and even more so as a place of education, potentially for their own children, interests that often coincided. Many of these were also VFR tourists, including those with children or other relatives at the University. Like other tourists their focus was on the Quad, but they had more time to spend, sought to see other buildings and wanted to see students in action.

> The campus is very beautiful. With nice lawn and buildings. But we did not learn much about the history of the university, which is a pity.

Others soaked up the atmosphere: 'The weather is not very good, but this building is so beautiful and looks just like an old castle. I so much envy the students in this university' while 'Walking in such a famous university is already a great honour, not to mention these beautiful and special buildings and the picturesque campus'. Many tourists noted that the University of Sydney was famous, with one woman (40–45 years) saying 'we always like to visit famous universities. They are educational icons of the country.' Another woman (25–30) said that the University of Sydney was famous, and that she wanted to 'to see the differences between the universities in Australia and in China' and had seen many photos of the University of Sydney before visiting the campus.

> I came here with my Mom because I plan to study abroad, and we want to see which country is better. The primary purpose of our trip is to learn more about the environment and atmosphere of the university, and the culture and history.

The greatest educational optimism came from a finance planner in their early 30s from Nanjing, but it was nonetheless indicative of the emphasis that some tourists placed on education:

> My son has just graduated from kindergarten. This is part of his graduation trip. I wanted him to see what universities look like. The University of Sydney is one of the best universities in Australia. I brought my son here to feel the atmosphere.

Some were thus participating in a 'strategic tourism' where future education was also linked to business and residential opportunities. Other responses were more straightforward:

> We know that Sydney University is very famous and has many strong subjects. The environment of the campus is better than I expected. Though seeing many tourists, I feel the study atmosphere is still good.

Similar to a number of other tourists, one woman liked the 'very English style' of the buildings, a view shared by three older women who commented that 'the building is European style, like a church'. One man (60–65) observed that 'the buildings are similar to the ones in Cambridge and Oxford' and another (25–30 years) noted that 'we feel a very different atmosphere on the campus from that of the campuses in our country'. Some tourists had visited the University several times, as repeat visitors, and three women (60–65) stated that they had come many times because it was spectacular.

Not surprisingly, the appreciation of heritage architecture and of the behaviour that was permitted was benchmarked against China. One woman (40–45) observed:

> I like this building. It gives the sense of history. I think one of the reasons why we Chinese people like this kind of buildings is because we have lost so many of our old buildings. Seeing such well-preserved old buildings stimulates admiration deep in our hearts.

Another female (35–40) remarked that 'the buildings are delicate with an intense sense of age. We can step on the lawn. This is one big difference.' The word *gulao*, loosely 'the quality of having heritage', was often used (Xie, 2016). Some were surprised that there was no official entry gate (as is usual in China), barriers or entry fees. Heritage, free access, views over the city, international educational strategies, peaceful open grassed spaces for relaxation, along with the Harry Potter myth, all combined in different ways to make the University of Sydney a 'marked place' for Chinese tourists.

The group tourists, the VFR tourists and many of those who came for educational reasons had learned (or not learned) of the university in a variety of ways, often through kin or 'general knowledge'. They were usually an older group. By contrast many of the younger tourists travelling independently identified the University as a 'check-in point' or 'marked place', with one female (under 25) saying, 'this building is now a marked place. I don't want to miss it'. Others responded similarly. Another female (30–35) commented that the University of Sydney is 'one of the four marked places. Also, we want our kids to feel the atmosphere of universities especially overseas universities. We strongly encourage our kids to study abroad when they grow up'. However, the significance of the University as a marked place was especially evident for younger, independent tourists, coinciding with those who were particularly interested in seeing the Harry Potter building.

The social media posts further demonstrate their role in contributing to the University becoming a marked place (in the absence of alternative marketing or guidebook coverage) and thus a tourist destination. They can be crudely divided into those that emphasised the significance of the university, its buildings and its campus and those that focused almost exclusively on the Harry Potter building. They were both descriptive and directive, and collectively offered interlocking rationales for visiting the University. We include a number of these, from Chinese platforms over a two-year period, to emphasise the various interrelated rationales for visiting the University. Most focused on heritage, architecture and the educational merits of the University. Inherent—and highly visible—in most posts was the photogenic nature of the university and the opportunity for taking photographs. Social media thus elaborated on the history and aesthetics (of buildings and landscape) that emphasised an established presence, sustained educational success, some link to the Chinese education system and an unusual cultural heritage.

VFR tourists were all independent tourists, often travelling with their relatives and seeing where they were studying, or had studied, with many contemplating the possibility of the University being an education destination for their own children. This cluster was most likely to want to see heritage buildings, actual students and the educational facilities. They indicated a desire to 'see the buildings, since my grandchildren may apply to Sydney University', and 'my younger sister may be applying to the University; we like the buildings and the atmosphere'. Other tourists reported similar objectives: 'If you go there you can easily combine this with the free Nicholson museum, the fish market or upbeat Newtown. I went there to visit my niece who studies there for her bachelor's degree. While walking around with her I found the main campus is just amazing. There are beautiful open spaces where you can sit and read a book or lie under the shade of a tree. It is a very pleasant environment which is ideal for doing studies' (TripAdvisor, 27 June 2017). Safety on the campus was never queried.

Educational futures and Harry Potter's past were often combined:

> For many, many years, since graduation, I never returned to campus. Because the little Harry Potter fans want to see the Harry Potter building, on this Sunday morning, I stepped back again. Looking at these once familiar buildings, I couldn't help but burst into tears. Go slowly and walk to the main building, where everyone is talking about the Harry Potter building. The blue cherry blossoms [jacaranda] can be seen everywhere on campus. I hope that in the future, [my son] Ajun can wear his master's gown to walk through here. (5 November 2019, *Mafengwo*)

For such returning students, a deep personal nostalgia was obvious, but that extended to others remembering their own education elsewhere and appreciating an atmosphere of learning.

The heritage buildings were a crucial backdrop. Many of those who came to see the Harry Potter building had been drawn by social media descriptions, and especially by photographs of the site. 'I learned from the internet that this building is one of the places where Harry Potter is filmed. The campus is very beautiful', 'Friends recommended me to come here to see the Harry Potter building'. Similar sentiments were replicated in the social media that also indicated overlapping interests and clusters: 'Sydney University is the first university in Australia. You should not miss this attraction when visiting Sydney. Not only to see the Victorian architecture, but you can feel the study environment of a foreign college, and even find the place where Harry Potter stayed' (iTrip, 28 March 2018). The focus could be quite narrow, and the Nicholson Museum (with the largest collection of antiquities in the southern hemisphere) was not mentioned to tourists, despite its being in the Quad, although, very exceptionally, social media had alerted visitors to its presence ('I also learnt that there is a museum that has mummies; I also want to see the mummies'). Many recommended spending time in the Quad to absorb the atmosphere and find the best locations and angles for photography (see Chap. 6).

Interest in the heritage buildings was often accompanied by an interest in seeing academics and students. By contrast, one concern of some Chinese tourists was the presence of other tourists. One female (50–55) tourist complained 'we see so many Asian people' and a female (30–35) said she saw 'so many Chinese tourists': a measure of the substantial numbers of Chinese tourists, recognition that there were few other tourists and disappointment at being unable to see students participating in classes. They could see the façade but not the workings of the University. Rationales, priorities and responses varied substantially, as tourists experienced and absorbed different facets of the place, partly influenced by age, time, inclination and other factors. None of these tours and visits were organised or directed by the University, nor were fees or restrictions imposed on tourists, leaving them to wander and spend as much time as and where they sought fit.

Finding Harry Potter

As various online posts indicated, there was a widespread belief that Sydney University and the Quad had a very close connection with Harry Potter. Half the visitors had come primarily for that reason. History and physical heritage were key parts of the touristic experience but were combined with the myth of Harry Potter.

> Dropped by when a relative was enrolled in the university. Beautiful facade and built a long time ago. Looks like Hogwarts. Here on a Sunday and many wedding couples here for a photoshoot (12 March 2019, *Mafengwo*)

For some, who had taken on board the notion that Harry Potter had a physical connection with the University, and that the Quad was Hogwarts, this was redolent of Hobsbawm and Ranger's (1983) 'invented tradition' embodying cultural and historical practices that claim continuity with a distant past but which are of recent origin, and here entirely invented. Not all tourists were however convinced by plausible falsehoods:

> The tour guide told us this is the Harry Potter building. However I know that Harry Potter is not filmed here but in Oxford. Even so, I understand why the building got this name because it does look like Hogwarts.

Both explicit and implicit in many posts is that the tourists had responded to suggestions on various platforms.

> When making my tour plan I saw pictures of this building on the internet. Many people recommend coming to take photos of this building. They also say this is the Harry Potter building.

> We came to see the Harry Potter building, The University is a famous marked place. It was very easy to find in trip advisor about the Harry Potter building. Many people recommend visiting.
>
> If you are a fan of Harry Potter, you must put this into your itinerary! Even if you are not, do spend a few hours there too. The building is ancient looking and very nice for photos (10 October 2019, *Mafengwo*)

Not everyone knew of the Harry Potter building. On group tours they were advised, in various ways, that the building was in some way linked to Harry Potter, while others had learned online. The more limited, and honest, explanation given by some guides was that the building closely resembled Hogwarts.

> We saw nice pictures of this building online, and we wanted to come and have a look. We heard that this building is called the Harry Potter building. I will recommend this place to my friends because it is a nice place to take photos. You know, tourists like taking photos

Even those who had not previously known of the Harry Potter building made comments such as:

> I didn't know about the Harry Potter building but now I will recommend it to others.

Consequently, over a very short time period, the University and its Harry Potter building became regular and positive presences on various platforms. In effect Chinese use of social media constructed a new site, focusing on a small part of the Quad and endowing it with a distinct but inaccurate heritage value. Tourism numbers grew, as it became a marked place through the creation of 'soft heritage' as the articulation of tangible architecture with cultural mythology.

That the University might be complicit in dubious marketing briefly became an issue early in 2019 when a spate of journalism and some Australian TV channels drew attention to the 'duplicity' and observed that promoting tours with inaccurate statements was in breach of Australian consumer law. The journalists also recorded typical statements from Chinese tourists:

> Our tour guide told us this is where one of the scenes of Harry Potter was filmed

> This place... Harry Potter... It's very famous so that's why I wanted to come here
>
> This place filmed or I think it filmed Harry Potter... it's very famous and I've seen looks of my classmates who have come here to take photos that are so beautiful so that's why I really wanted to come here (Evlin, 2019)

However, no tourists interviewed by journalists, or in our interviews, expressed concern that this was not the 'real' Hogwarts and nothing subsequently changed. They had negotiated any ambiguities and expectations. While experiencing a place vicariously, where the location presented in the films was both elsewhere and artificial, all enjoyed the experience, as many recorded online:

> Our purpose in visiting Sydney Uni is very clear, for the Harry Potter Building. But I know Harry Potter is actually not filmed here. This is a beautiful misunderstanding. This building is definitely a spot you should see when visiting Sydney. (4 January 2019, *Mafengwo*)

> Someone lying on the lawn of the school was reading a book, and the lake was filled with ducks and unfamiliar birds. 'Harry Potter House' is the main teaching building of the University of Sydney. The first flight lesson of Harry Potter and the classmates was on this big lawn! (9 December 2018, *Mafengwo*)

For all tourists, the Quad was the epicentre of tours and the site for photographs, especially selfies (Fig. 5.2). Even those primarily interested in future education rarely went further than the Quad, where there were several older but functional lecture theatres.

While it is impossible now to determine where and how the myth originated, and whether the myth of Harry Potter and Hogwarts was what first drew tourists to the University, the genesis of Harry Potter tourism occurred at much the same time that a distinguished Chinese star of popular culture had her wedding photographs taken in the Quad, with the same backdrop eventually used by countless other Chinese tourists (see Chap. 6). Photography was absolutely critical in posts and in establishing the Quad as a tourist site.

What amounts to informal marketing by amateur influencers through Little Red Book, *Mafengwo* (and other social media) both in China and in Sydney, above all through diffusing pictures of the Quad, was highly successful. That occurred in a context where there had been no marketing of

Fig. 5.2 Harry Potter tourist in the Quad

the Quad as a physical or heritage structure (although the University marketed itself educationally to overseas students) or of the prestigious Nicholson Museum. Knowledge of the Harry Potter building can only have come from informal social media channels (or actual word-of-mouth, though no-one, including VFR tourists, claimed that as a source). Indirectly Harry Potter is likely to have boosted the number of Chinese international students in the University. While it is impossible to correlate enthusiasm on social media with actual subsequent visits, the number of such statements and expressions of intent, suggest the very strong probability that the Quad as a tourist destination could only have been created by social media.

Fans and Followers

As knowledge of Harry Potter's aethereal presence at Sydney University increased so too did the numbers of Harry Potter fans, distinct from earlier tourists who simply enjoyed the similarities of the Quad and Hogwarts, and were able to get in the mood. The 'real' fans had come specifically for a Harry Potter experience, with the enthusiasm and passion that constituted a form of secular pilgrimage, even a search for or validation of personal meanings. That could occur in a short time in an accessible, elegant and peaceful place even amongst 'non-believers'. Performing, photographing and imagining do not take long: very much a niche experience, typifying the incidental nature of media and cultural tourism as often a part of something broader.

Unlike many other popular sites linked to popular culture, and well-known on social media, there are no signs, boundaries or admission fees in the Quad, making this a particularly intimate site. It is therefore quite unlike the most famous Harry Potter site, Platform 9¾ at Kings Cross station in London, or even the Wizarding World of Harry Potter (WWOHP) in Orlando, Florida, both of which are official and commercial (Larsen, 2015; Milazzo & Santos, 2022; Waysdorf & Reijnders, 2018a, b). The lack of any signage and commercialisation nonetheless provides a degree of authenticity and freedom for fans alone. They are neither directed nor disciplined in any way and are free to imagine. Accessing the site via social media directions and suggestions to some extent guarantees that the site will be 'authentic, unique and real, especially for those who consider themselves true fans and are eager to experience the authenticity of the destination' (Segota, 2018, p. 431). Authenticity is thus in the eye of the beholder. An imaginary Harry Potter and Hogwarts are reimagined in Australia.

Without the imagination of ardent fans, and the role of social media, the Quad would hold only architectural interest for tourists from another country. This is changing as more Chinese students graduate from the University, making the site important as a personal milestone in the life of some Chinese people. Its wider cultural significance has been entirely shaped and curated by Harry Potter fans, but as a 'sacred site' for fans only. As with Dracula fans at the Bran Castle (Romania) (Banyai, 2010), fans at the University of Sydney find authenticity in Harry Potter connections, rather than in the history or architecture of the Quad. Somewhat similarly, Nördlingen in Germany, a tourist destination for Japanese fans of a manga franchise called *Attack on Titan*, who had also followed social media directions and suggestions, was unknown to local people, and Nördlingen was officially denied as a site related to *Attack on Titan*. The absence of commodification there, as at the University of Sydney, effectively demanded that the 'fan tourists' independently performed their own activities at the site, creating a 'deeper involvement with the site through performing authentic experiences of fan pilgrimage' so developing imagination and affection for the site (Thelen & Kim, 2023, p. 16) while sharing its significance for those who might follow, so enhancing a sense of fandom and community. However, at the University of Sydney, unlike Nördlingen and other more esoteric media sites, including several overseas anime sites (Jang, 2020; Jang & Yamamura, 2020), some film tourist sites (Tan, 2024), and art and music sites (Gibson & Connell, 2005), other tourists shared and enjoyed the possibilities (and performances).

Managing the Harry Potter Experience

Management of campus tourism at the University of Sydney was almost entirely passive. Other than banning tour buses from the campus, there was no concerted attempt to manage, regulate, encourage or discourage Chinese (or any other) tourism, charge visitors or market merchandise. Commercial outlets on campus had little to sell of great interest to tourists (and nothing that hinted at Harry Potter). Heritage tours were few, although visitors seeking to have wedding photos taken on campus (a frequent activity) are charged for the privilege.

Tourism may have negative impacts, as it did in China, if tourists encroach on lectures, offices or libraries physically or aurally—a situation becoming a cause of concern by the mid-2010s. That was largely avoided at the University of Sydney, mostly fortuitously, by the relative absence of lectures in the Quad itself, but it offered a challenge. (One of us, who regularly lectured in the Quad, became familiar with tourists appearing at the back of the lecture theatre during lectures). Disruption to people working in offices in the Quad was incidentally solved mainly by moving many senior administrative functions, including the Vice-Chancellor, to a new building. Vehicle congestion was resolved in 2016 by the relocation of tour buses to a suburban street within walking distance. Other actions included the erection of signs (albeit in English) politely reminding everyone to minimise noise in a working environment. Similar strategies have proved essential in other universities (Westfall, 2017). Tacit acceptance and even support were achieved through detachment and tolerance.

The management of campus tourism was thus driven by reaction rather than pro-action: the desire to overcome such negative aspects as transport congestion and the disruption of academic and administrative activities. Any interest in creating a strategic tourist vision, and learning from other educational institutions, dissipated once immediate issues were resolved. For some it remained a mystery; a very distinguished university professor, with an office close to the Quad, asked at a 2016 campus management meeting, 'Do we know why the Asian tourist market is so interested in the University? If we knew when they were coming we would have a better way of engaging with them'. Despite being almost entirely reactive to the rise in Chinese-generated campus tourism, the University maintained free and public access to its campus, met security concerns and requested appropriate behaviour. This encouraged more tour companies to include the campus among their sites, as a cost-free destination, both as an inner-city location for tours that stay close to the city centre, and as a stopping point (aided by the presence of toilets) for longer tours. Only incidentally therefore, through minimal but necessary site management, did the University play even an indirect role in stimulating the growth of campus tourism. This was partly intentional by the University of Sydney, which, like leading universities in China, was understandably focused on education rather than tourist experiences.

Curiously, senior university bureaucrats initially denied any knowledge of the notion that the Quad might be similar to Hogwarts and claimed that the growing numbers of tourists were mainly there to visit the fine sandstone Gothic buildings and to visually assess the potential for future educational opportunities. They were however concerned that the tourists were sometimes noisy and disruptive of sensitive intellectual matters (especially when the Vice-Chancellor himself had an office in the Quad). At the same time they were secretly fascinated and delighted by the interest in the University, however unusual it was. When questioned, the management argued that the University had no interest in promoting more tourism but were only concerned to regulate excessive noise and any disruption of standard university activities.

Simultaneously, and somewhat clandestinely, the University did engage in subtle forms of place marketing. Counteracting notions that the University was uninterested in tourism, and even that it disavowed any notion of Harry Potter, it announced in 2017 on its Facebook page that

Fig. 5.3 The 'Hogwarts crest' outside the Great Hall, University of Sydney

the pillars of the most famous part of the Quad had 'been given a 21st century makeover, of the magic variety, thanks to the creativity of an imaginative stonemason', with a new crest located on a previously blank sandstone shield on the University of Sydney's Great Hall in the Quad (Fig. 5.3). While several shields have been etched with various distinguished local family crests over the years, and some remain blank, it marked a considerable move away from the traditional style: the school crest of the Hogwarts School of Witchcraft and Wizardry. That the announcement appeared on April Fool's Day suggested that the University rather enjoyed its newfound fame; its 'stonemason' queried, 'Who knows, maybe one day Sydney Uni might even offer a Bachelor of Dark Arts or PhD in Muggle Studies'. Sometimes the imaginary is based on myth, such as Harry Potter and Hogwarts as literary inventiveness, but the sandstone and gargoyles of the Quad provided a place for the myth to be materialised.

By then disavowal of the aethereal presence of Harry Potter was impossible. In the same year, BuzzFeed, the American Internet media, news and entertainment company, with a focus on digital media had ranked 'University Campuses That Might Secretly Be Hogwarts' (www.sydney.edu.au/news-opinion/news/2017/03/31/great-hall-gets-a-magic-makeover.html) to proclaim that the University of Sydney's Quad was the closest approximation to Hogwarts (only out-ranked by Oxford University where some scenes from the first two Harry Potter films had been filmed). That may have been the first time that a wider public had been alerted to the parallels between the University of Sydney and Hogwarts. Perhaps too the University's pretence that 'there was nothing to see here' now required a response to the growing numbers of tourists and at least a quiet recognition that the University had become a tourist site.

With heightened global Harry Potter fervour, two decades after the first book had appeared, other organisations produced similar lists and stories. Study International ranked universities where student might dream of studying in Hogwarts-like buildings. After several American universities, Sydney University was ranked a disappointing tenth: yet it was 'Perfect for Quidditch Practice, Wouldn't You Say?'. After all, 'The Great Hall at Sydney University looks exactly like Hogwarts' Great Hall. It's even got a quadrangle where students can sit and read spell books. Or just normal books, I guess'. The myth was becoming increasingly global.

The Market Discovers Harry Potter

If the University was reluctant to admit any formal links with Harry Potter, despite the new crest, but, as the reportage by Buzzfeed and Study International suggests, the commercial world was becoming conscious of its economic potential, as it emerged in standard guidebooks. Inclusion only occurred in response to the visibility of Chinese tourism. Indeed, with this distinctive exception, universities were not otherwise perceived as tourist attractions, even in Australia where nineteenth century heritage is rare, and Gothic sandstone buildings even rarer. Moreover, the University of Sydney is barely a couple of kilometres from the city centre.

By way of contrast, Chinese-based social media, such as *Mafengwo* and Little Red Book, highlighted the role of tourists sharing images and experiences in real time on almost any device and platform (Sigala, 2016) and by doing so encouraging similar tourists to visit the university. Despite a more complex rationale, this process thus paralleled, in a very different country, the growth in campus tourism in China (Chenyu, 2018; Mangi et al., 2019). Trust in social media when people of similar backgrounds, typified as influencers, recommended places and experiences, offered directions and thus stimulated (and effectively created) the evolving development of campus tourism in Sydney, beyond the passive experience of traditionally stereotypical group tours, to become a 'marked place'. The University of Sydney responded to better manage the site, rather than be actively involved in market co-creation. Fortuitously, the relocation of tour buses just off-campus enhanced the heritage architecture, enabling superior photographs of the spectacle of the Quad. In an era of sharing images on social media, this proved important for the tourist experience and for the branding of the University, although that was not the objective.

Both the Chinese entrepreneurs who ran the package tours and the more independent Chinese tourists themselves, through eWOM, made the University of Sydney a distinctive tourist destination, and shaped and reshaped the experience. The independent tourists went substantially beyond the more passive gaze of the package tourists, at least in interactions with others in the Quad, the active performance of photography (using magic wands, seeking to fly and wearing colourful scarves and capes) and the transmission and circulation of the images (and assorted suggestions and stories) to stimulate others. Their posts indicated that this was invariably a positive process.

Chinese tourists had placed the University on the tourist map. The private sector stepped in. Just as the University realised that there were possible advantages in discretely acknowledging the presence on campus of Harry Potter and Hogwarts, and guidebooks took note, so the wider tourist industry too grasped the possibilities. The arrival of COVID-19 nipped interest in the bud, but by early 2023, before the return of Chinese tourists, prospects of local markets had appeared. Ella Tours thus advertised the Harry Potter High Tea:

> Named the Wizardry High Tea, this spellbinding journey boasts interactive characters, plenty of activities, a lot of butterbeer for everyone, and wicked polyjuice potions and cursed cocktails for the twilight (18+) iteration of the event! Taking place at the magical halls of Sydney Uni, the Wizardry High Tea will have three sessions per day—11 am and 2.30 pm for all ages, and

> 7 pm for adults only. Bookings are essential. Join us at our spellbinding High Tea! Christmas in July Edition. Enjoy an enchanted high tea, interactive roaming characters for the family, Wizardry Activities with a Yulefest twist. Hidden surprises and more....Our twilight guests will be treated to wicked Cursed Cocktails! In addition please do check out A Wizardry Day Out a new family day out for all magical beings!

Other tour companies also sought a share in the domestic Potter market. Officer Travels were marketing Sydney's Very Own Hogwarts early in 2023, with a message liberally illustrated with pictures of the Quad:

> Where to find magic in the heart of Sydney. Did you know that Sydney had a building very similar to the one used in the film. While the University of Sydney Harry Potter location isn't an official film set, it's so similar that you'll feel like real magic is at play. Long gardens and winding paths lead up to the grand entrance but the real magic happens when you step through the archway and into the courtyard. Once inside the courtyard it's easy to imagine Mud bloods and wizards walking around here donned with cape and wands, fighting off trolls, playing Quidditch. It's no wonder the Muggles here have set up their own Quidditch teams. This building's fame goes back beyond its magical roots; it is Australia's first university dating right back to 1850.

In the 2023 midwinter break at the University, Ella's List was offering a Magical Harry Potter High Tea: 'taking place in 'the magical halls of Sydney Uni, the Wizardry High Tea will have three sessions per day ...for all ages'. There was now no question that the University had embraced Harry Potter, albeit for a domestic market, even before Chinese tourists had returned. It was a far cry from previous variants of campus tourism.

Whose Campus Tourism?

The Chinese tourists at the University created their own version of campus tourism (while a much smaller largely invisible group of local people undertook limited guided tours of the University, more redolent of campus tourism in other countries). By sharing words and images through various social media, Chinese tourists influenced other tourists, without any role being played by the University, rather different then from most tourism sites (such as cathedrals, that have been quicker to recognise the potential to promote tourism and collect donations when observances are not taking place) and despite there being potential advantages to be gained from promotion and marketing. Universities have sought to avoid 'cathedral parallels'—prioritising education over tourism—and merely tolerated and 'managed' tourism: circumstances vastly different from any 'fiercely competitive experience economy' (Chen et al., 2020b, p. 1408). That experience and the possibility of profitability were taken up belatedly.

Both tour groups and individuals visit Sydney University as it has become a 'marked place' significant enough to have belatedly filtered back into a standard guidebook to Sydney. Despite considerable interest in attracting international students, the University has not forcefully advertised, promoted or run tours (other than aimed at prospective students) and has effectively remained somewhat distanced from the city itself and the tourists that flow through (McManus et al., 2021). University campus tourism remained institutionally underdeveloped while at the Quad, just as at other 'new' Chinese tourism sites (see Chap. 6), there is minimal commercial presence and little for tourists to buy. The quest for Harry Potter caught the University unawares.

Chinese tourists experienced the University environment as 'peaceful', spacious and 'heritage'. However the group tourists were more likely to be critical than those who had chosen to visit the University ('I feel this part of the university is exploited to be a tourist attraction. Where do students have their classes?'). They were also more likely to be given doubtful information ('Our tour guide just told us this was the Harry Potter building'). Unencumbered by rules about visiting hours, or places they could or could not go, the tourists experienced perhaps the quietest part of a 'working campus'. They followed suggestions in social media and, through social media, they subsequently created and embellished this as a tourist experience for other Chinese tourists. Effectively they shaped it into what they wanted it to be—a place that offered heritage, learning, tranquillity and Harry Potter mythology; some embraced the entirety of this, others less so. Shorn of Harry Potter, Chinese campus tourism was remarkably similar to that in Korea and Thailand, with an interest in nostalgia, prestige, heritage architecture and an educational ethos (O'Regan & Choe, 2022), and more broadly to that in European and American universities.

Education, and contemplating a future for children, has long produced a distinct niche in the tourist industry, prominent in the northern hemisphere, but which spread from China to Australia in an unusual form. That illustrated how niches were rarely 'pure' but had mixed objectives, changing over time and within family groups, and represented a rare variant of the tourism industry that could succeed without formal marketing. Unless tourists also travelled to other universities (and there was no indication that this was so), then the niche lasted no more than part of a day within a longer tour, and took in only one of Sydney's six universities. That suggested that for many visitors at least part of the rationale centred on Harry Potter and heritage buildings (that no other Sydney university could emulate). Although eWOM social media guidance to those yet to visit suggested that they stay for several hours, relax and take many pictures, and many followed that advice, once again this was a niche that was unlikely to extend more than a few hours let alone an entire tour. Even a triple objective (education, heritage architecture and Harry Potter) could easily be achieved in a day. Indeed many tourism niches—certainly where they involve specific sites—are probably often the highlights (or perhaps occasionally the disappointments) of a much longer tour, rather than the objective of an entire vacation.

Since visitors went to no other universities, and there are no other Harry Potter sites in Australia (nor any comparable Gothic buildings), it is therefore reasonable to ask whether tourism at the University of Sydney was simply a distinct and pleasurable conjuncture of factors, and convenient for many, as a highlight of a 'standard' tourist experience. However, while interlocking interests and experiences had begun to make the University something of a tourist icon, for some it was an essential ***daka*** site. The University had found an accidental and dual niche in the tourist industry.

Following Harry Potter

Since the mid-2010s, the myth of Harry Potter has remained strong. Until recently the University of Sydney neither figured in guidebooks or attracted tourists, and its touristic prominence has been entirely due to the role of Chinese tourists functioning as influencers. While that influence has not displaced other 'marked places', such as the Opera House, it has provided an intriguing addition, so simultaneously complementing and usurping traditional tourism media. The emergence of the University and other new sites provides a distinctive example of how incidental 'online brand advocates', or influencers—usually simply fellow tourists—can restructure the nature of tourism from below, a theme we take further in the next chapter.

Evolving Chinese international tourism at Sydney University vividly demonstrates the widespread usage and impact of social media, where rapid economic, technological, and societal changes, and depictions and descriptions of particular sites, have enabled shifts in travel and tourism-related mobility (Xu & Wu, 2016), here creating a new and distinctive destination. The hitherto unrecognised and newly invented association between architectural heritage and Hogwarts spectacularly gave rise to Harry Potter, placing a new site in the ranks of popular culture destinations. The simultaneous rise of social media and independent travel has shaped new patterns of tourism, distinct from established forms, where social media are both a key source of travel information and inspiration and a demonstration of middle-class practice and performance. Social media can create and boost particular sites through mediated self-presentation. Tourists share accounts and experiences, reinforced by others and trusted by fellow citizens, where language and visual representations are important, as social media support and steer, but also divert from, the local evolution of tourism. This is unusual and distinctive in being oriented to and taken up by many people who have already embarked on a tour and are commenting on and modifying its structure whilst en route. Online blogs and social media—with smartphones and photography at the core—have shaped a new form of tourism increasingly divorced from package tours and guidebooks but centred on new social media platforms (such as Little Red Book) that offered eWOM recommendations and directions. Their combined significance has been such that becoming a marked place, as has occurred at the University of Sydney, can generate substantial tourist numbers, as Chinese tourists shape the motivations, preferences, and interpretations of likeminded others. Trust in the recommendations and observations of those of the same national origin and social status, and using the same internet platforms is invariably positive and recognises that posts on these platforms are likely to be credible, recent and relevant and, of course, can be shared instantaneously. On-site activities and the post-visit phases of tourism have become blurred and merged; there is a considerable difference from tourism in earlier decades when communication of experiences and images involved post-cards, printed photographs and lengthy delays. Social media have made the printed word largely irrelevant. For most independent Chinese travellers their compatriots are the primary source of information.

The 'consumer networks' created through social media enabled tourists to better discriminate, and benefit from new, up to date and more comprehensive knowledge, enabling seeking out and discovering sites and experiences of personal interest, pleasure and relevance. That has structured the University as a place for assessing education opportunities, vicariously experiencing Harry Potter, enjoying heritage architecture and buildings, and taking photographs, activities that usually overlap. The Quad has become a new niche in literary and film tourism, a place of imagined authenticity and of soft heritage. Independent tourists have created a new, idiosyncratic and more personal tourist space, where aesthetics and popular culture are combined, linked in to both major attractions and other alternative sites across the city. It is these other, even more recent, sites in Sydney and beyond, that we turn to in the next chapter, where the significance of influencers, photography, gender and colour become more obviously important; the role of the tourists themselves in creating sites becomes even more evident. The University of Sydney proved simply a prominent and intriguing precursor to other more widespread changes.

Bibliography

Albino, S. (2015). *Tourism in University Cities: The role of universities in place branding*. PhD thesis, University of Exeter and Universidade de Lisboa. https://ore.exeter.ac.uk/repository/handle/10871/21034

Banyai, M. (2010). Dracula's image in tourism: Western bloggers versus tour guides. *European Journal of Tourism Research, 3*(1), 5–22.

Beeton, S. (2016). *Film-induced tourism* (2nd ed.). Channel View.

Bronner, F., & de Hoog, R. (2011). Vacationers and eWOM: Who posts and why, where, and what? *Journal of Travel Research, 50*(1), 15–26.

Chen, X., Cheng, Z., & Kim, G. (2020b, 1904). Make it memorable: Tourism experience, fun, recommendation and revisit intentions of Chinese outbound tourists. *Sustainability, 12*(5).

Chenyu, L. (2018, July 6). Top Chinese universities cap visitors as campus tourism soars: Peking and Tsinghua universities will only allow a few thousand daily guests as a crowd control measure. *Sixth Tone*. Accessed October 28, 2019, from https://www.sixthtone.com/news/1002576/top-chinese-universities-cap-visitors-as-campus-tourism-soars#

Connell, J. (Joanne). (1996). A study of tourism on university campus sites. *Tourism Management, 17*(7), 541–544.

Connell, J. (Joanne). (2000). The role of tourism in the socially responsible university. *Current Issues in Tourism, 3*(1), 1–19.

Connell, J., & Rugendyke, B. (2010). Creating an authentic tourist site? The Australian standing stones, Glen Innes. *Australian Geographer, 41*(1), 87–100.

Erni, J. (2008). Enchanted: Harry Potter and magical capitalism in urban China. *Chinese Journal of Communication, 1*(2), 138–155.
Erni, J. (2013). When Chinese youth meet Harry Potter. Translating consumption and middle-class identification. In A. Fung (Ed.), *Asian popular culture: The global (dis)continuity* (pp. 21–41). Routledge.
Evlin, L. (2019, March 4). The 'magical' reason Chinese tourists are flocking to the University of Sydney. *SBS News.*
Gibson, C., & Connell, J. (2005). *Music and tourism. On the road again.* Channel View.
Gothie, S. (2016). Playing 'Anne': Red braids, Green Gables, and literary tourists on Prince Edward Island. *Tourist Studies, 16*(4), 405–421.
Hayward, T. (2020, July 4–5) Cambridge's new tourist problem. *Financial Times,* 14.
Hobsbawm, E., & Ranger, T. (1983). *The invention of tradition.* Cambridge University Press.
Hwang, Y. (2011). A theory of unplanned travel decisions: Implications for modeling on-the-go travelers. *Information Technology and Tourism, 12,* 283–296.
Jang, K. (2020). Creating the sacred places of pop culture in the age of mobility: Fan pilgrimages and authenticity through performance. *Journal of Tourism and Cultural Change, 18*(1), 42–57.
Jang, K., & Yamamura, T. (2020). Creative fandoms and the mediatized sacred sites. In M. Månsson, A. Buchmann, C. Cassinger, & L. Eskilsson (Eds.), *The Routledge companion to media and tourism* (pp. 307–315). Routledge.
Larsen, K. (2015). (Re)claiming Harry Potter fan pilgrimage sites. In L. Brenner (Ed.), *Playing Harry Potter: Essays and interviews on fandom and performance* (pp. 38–54). McFarland.
Le Heron, E. (2020). Assembling film-landscape relations: Performative practices of The Lord of the Rings tours. *New Zealand Geographer, 76*(1), 39–48.
Lee, C. (2012). 'Have magic, will travel'. Tourism and Harry Potter's United (Magical) Kingdom. *Tourist Studies, 12*(1), 52–69.
Lexhagen, M., Larson, M., & Lundberg, C. (2013). The virtual fan(g) community: Social media and pop culture tourism. In *Tourism social media: Transformations in identity, community and culture* (pp. 133–157). Emerald.
Li, W. (2013) *Research on the development of campus tourism – Taking the universities of China and Britain as an example.* PhD thesis, Nanchang University.
Ling, L., & Dai, Y. (2012). Discussion on the sustainable development of campus tourism. *Market Forum, 6,* 94–95.
Lonely Planet. (2017). *New South Wales.* Lonely Planet.
Mangi, M., Yue, Z., & Kalwar, S. (2019). Universities emerging as tourism sites in China: A case study of Tsinghua University Beijing. *Cogent Social Sciences, 5,* 1–26.

Mannheimer, E., Reijnders, S., & Brandellero, A. (2022). Game of Thrones tourism and the (re)imagination of the new Northern Ireland. *International Journal of Cultural Studies, 25*(2), 554–569.

McManus, P., Connell, J., & Ding, X. (2021). Chinese tourists at the University of Sydney: Constraints to co-creating Campus Tourism? *Current Issues in Tourism, 24*(24), 3508–3518.

McWha, M., & Beeton, S. (2018). Engaging with popular media through tourism: A brief introduction. *Tourist Studies, 18*(3), 257–260.

Milazzo, L., & Santos, C. (2022). Fanship and imagination: The transformation of everyday spaces into *Lieux D'Imagination. Annals of Tourism Research, 94*, 103399.

O'Regan, M., & Choe, J. (2022). Searching for prestige: Motivations and managerial implications of Chinese campus tourists. *Leisure Studies, 41*(6), 862–878.

Rough Guides. (2017). *The Rough Guide to Australia* (12th ed.). Rough Guide.

Segota, T. (2018). (G)A(i)ming at the throne: Social media and the use of visitor-generated content in destination marketing. In C. Lundberg & V. Ziakas (Eds.), *The Routledge handbook of popular culture and tourism* (pp. 427–438). Routledge.

Sigala, M. (2016). Social media and the co-creation of tourism experiences. In M. Sotiriadis & D. Gursoy (Eds.), *The handbook of managing and marketing tourism experiences* (pp. 85–111). Routledge.

Spracklen, K., & Spracklen, B. (2014). The strange and spooky battle over bats and black dresses: The commodification of Whitby Goth Weekend and the loss of a subculture. *Tourist Studies, 14*(1), 86–02.

Symington, A. (2018). *Sydney.* Lonely Planet.

Tan, Y. (2024). 'Capturing the shining moments': Performing liminoid Seichi Junrei on social media. *Current Issues in Tourism, 27*(12), 1959–1972.

Thelen, T., & Kim, S. (2023). Understanding fan tourists at a non-commodified fan pilgrimage site: An application of fan capital theory. *Current Issues in Tourism, 26*(1), 12–17.

Tung, V., Lee, S., & Hudson, S. (2017). The potential of anime for destination marketing: Fantasies, otaku, and the kidult segment. *Current Issues in Tourism, 22*(12), 1423–1436.

Waysdorf, A., & Reijnders, S. (2018a). The role of imagination in the film tourism experience: The case of Game of Thrones. *Participations. Journal of Audience and Reception Studies, 14*(1), 170–191.

Waysdorf, A., & Reijnders, S. (2018b). Immersion, authenticity and the theme park as social space: Experiencing the wizarding world of Harry Potter. *International Journal of Cultural Studies, 21*(2), 173–188.

Westfall, S. (2017, October 13). Next stop... Yale: Chinese tourism on campus. *The Politic.* https://thepolitic.org/next-stop-yale-chinese-tourism-on-campus/

Wong, J., Lai, I., & Tao, Z. (2020). Sharing memorable tourism experiences on mobile social media and how it influences further travel decisions. *Current Issues in Tourism, 23*(14), 1773–1787.

Woodward, S. (2013). Campus tourism, universities and destination development. In I. M. Smith & G. Richards (Eds.), *The Routledge handbook of cultural tourism* (pp. 265–272). Routledge.

Xie, L. (2016, August 23). How nice is the quad? *Honi Soit.*

Xu, H., & Wu, Y. (2016). Lifestyle mobility in China: Context, perspective and prospects. *Mobilities, 11*(4), 509–520.

Yousaf, S., & Fan, X. (2020). Copysites/duplitectures as tourist attractions: An exploratory study on experiences of Chinese tourists at replicas of foreign architectural landmarks in China. *Tourism Management, 81*, 104179.

Yuming, Y. (2018, July 6). Personal summer tour to Peking University Tsinghua needs online reservation. *Beijing News.* http://www.bjnews.com.cn/feature/2018/07/06/493977.html

Zhang, L. (2017, August 23). 430,000 people visited Peking University, Tsinghua University during the summer, 30 of whom were detained for disturbing order. *The Legal Evening News.* https://www.thepaper.cn/newsDetail_forward_1772573

Zhang, X., & Ryan, C. (2018). Investigating tourists' and local residents' perceptions of a Chinese film site. *Tourist Studies, 18*(3), 275–291.

Zhou, F. (2009). Analysis of the current development of campus tourism and counter measures. *Contemporary Education Theory and Practice, 1*(5), 154–156.

CHAPTER 6

Daka: A New Chinese Tourist Geography

This chapter examines the increasingly important role of photography in creating and representing the various sites visited by Chinese tourists. Photography created and enhanced new sites for the specific gaze of other Chinese tourists in Australia, and relatives, friends and potential tourists in China. Smartphone photography has modified and reshaped a tourist gaze, present in the past through 'happy snaps' but restructured here by a millennial generation, with gender implications. While tourists have always shared photographs and videos of their holiday destinations, with themselves in place, tourism and photography being inseparable (e.g. Tussyadiah & Fesenmaier, 2009), in the second half of the 2010s it took on new forms in the hands of 'amateur influencers', typified by *guimi* ('best girlfriend' or 'girlfriend getaway') tourists, who combined fashion, colourful locations and photography. In some respects that emanated from the experiences and performances of Harry Potter fans in the University of Sydney Quad.

The expanding use of smartphones as mobile media offered 'new embodied, sensory, corporeal and emotional experiences [that] are increasingly engaging and entangling with contemporary ways of living, working and playing' (Lin & Yang, 2021, p. 677). That richness was evident in how smartphone use enabled a ubiquitous connectivity creating a new type of indirect and mediated interaction at a distance: even an 'ambient co-presence' (Madionou, 2016, p. 183). That then spawned a new form of 'instant tourism' centred on the instantaneous availability of attractive

J. Connell et al., *Chinese Tourism in Australia*,
https://doi.org/10.1007/978-981-97-2477-2_6

visual imagery and, consequently, the equally instantaneous ability of independent tourists to restructure travel with on-site decision making.

This chapter continues and develops the central theme of the previous chapter to discuss how the use of Chinese social media, notably Little Red Book or Red (*xiaohongshu*)—a more focused version of WeChat—effectively created multiple new sites of interest to Chinese tourists and which amounted to new travel itineraries. These were places not usually visited by Australian residents or tourists from other countries, hence their elaboration produced a new and dynamic Chinese tourist geography grafted on to established tourist and cultural geographies of Sydney. This geography was based around the idea of a 'marked' or 'check-in' place (网红打卡点) (daka), where '网红' means internet star or instant online celebrity, that should be visited by other Chinese tourists.

Daka tourism emerged in China in the late 2010s as an internet phenomenon dominated by millennials, and generated through the internet, as tourists followed suggestions and directions posted on social media. Large numbers of tourists followed, often to precisely the same spot to take photographs in exactly the same place, in what has been described as 'the ritualistic uploading of a picture or video from a photogenic spot by tourists to social media' (Tan & Yang, 2021, p. 3489). A considerable majority of such *daka* tourists are middle-class women in white collar employment, including students, and a majority with university degrees, travelling with friends or relatives (Ni, 2020). The website Douyin became hugely popular and featured such categories as 'food' and 'scenic sites'. Something of a sub-culture developed, with the emergence in China of *daka zu* (*daka* tribes), so that, as particular places (and also particular restaurants) became suddenly fashionable, and numbers increased, congestion often occurred. Fashionability could be linked to a media presence, such as the city of Zhanjiang, that became a destination after being on a hit TV show, *The Bad Kids*. Some places, like a coffee shop in Beijing built inside a shipping container, sought to make themselves as *daka*-friendly as possible, while others, such as a bus-station in Hangzhou were caught unawares, generating a two-hour long queue to photograph it to demonstrate that one was there (Anon, 2019; Ni, 2020). Sites may therefore be quite ephemeral, temporary 'hot destinations', and visits fleeting.

Those who shaped the new geography of Sydney were not celebrities, nor were they only interested in ticking-off sites, but were more engaged in photography. Neither did their recommendations generate crowds. Ironically many recommendations advised how to avoid crowds at popular sites. Since almost all users of *xiaohongshu* were young women, the new geography was distinctly gendered. Photo-taking and sharing have become ubiquitous in emerging middle-class Chinese society and 'selfies' ('self-shot' or *zipai*), loosely considered here to be photographs where the photographer, or a friend, is a key part of the picture, have become an enthusiastic and everyday performative and social practice, and thus embedded in contemporary Chinese touristic practices (Kimber et al., 2019; Liu, 2021). In these contexts, 'the perfect "selfie" [is] summarised as the selfie taken at an internet-famous location, showing an attractive self or an online celebrity style' (Liu, 2021, p. 245). Hitherto, the importance of place in performativity, photo taking and posting on social media has scarcely been identified and analysed, despite a lineage of studies that have followed from observations of tourists at the Taj Mahal in India (Edensor, 1998) and the recognition of popular culture sites (Chap. 5) created primarily by the tourists themselves.

Tourism is primarily a visual experience, long recorded by postcards from and posed photographs at tourist sites. Such photographs and postcards emphasised a presence there but were largely static and far from instantaneous. As smartphones and social media have emerged in recent years the practices of photography have changed with images—often variants of selfies—being instantly transmissible not only to kin and friends but far beyond (both socially and geographically). That has enabled messages to be combined with photographs: a contemporary postcard conveying a more personalised account of the destination.

Scarcely surprisingly just as postcards usually came from colourful locations so photographs were taken at colourful sites. Contemporary Chinese tourists embellished this by deliberately seeking out places that would make elegant backdrops to photographs of themselves, especially for wedding photographs. These sites were invariably photogenic places, such as Diamond Bay and Lavender Bay in Sydney, and various lakes, colourful fields and buildings elsewhere. Part of the attraction of particular places related to colour, and the focus on colour extended from the sites themselves to the need for tourists to display themselves (so engaging in forms of performativity) and photograph themselves at the site (and be encouraged to dress appropriately, according to the nature of the site, season and time). The quest for fashion and colour, whether designed, painted or natural, was guided by and principally directed at other actual or potential female tourists. Smartphones enabled tourists to be more spontaneous and creative, and achieve more satisfaction, and to post information, photographs and enthusiasm, to friends, relatives and multiple others, so becoming instantaneous influencers to a few or many. Despite the established capacity for tourists to make on-site decisions over travel, smartphones raised this to a new level so changing the travel context (Hwang, 2011; Liu et al., 2022), and creating new destinations. The democratic and participatory character of social media has directly increased the role of mainly millennial tourists in place-making.

Capturing and Being the Visual

Colour and emotional connections with sites became key elements of Chinese tourist distinctiveness and place-making. More passive gazing was inevitably extended into active performing, as elsewhere (Perkins & Thorns, 2001; Larsen, 2005), and performativity became more elaborate as photographic technology (notably videos and thus vlogs) became more complex. Photography encouraged performativity. At the Quad and elsewhere, tourists spent 'considerable amounts of time striking poses that are related to their image of self, their dreams and achievements: poses that indicate how they want to be seen, with whom, and what they are proud to have seen' (Stylianou-Lambert, 2012). At the popular destination for independent Chinese tourists of Pai (northern Thailand), enterprising local people constructed colourful attractions, such as a collection of giant red strawberries, a 'Swiss cottage' and an upside-down house, to be incentives for tourism and backgrounds for photography (Kimber et al., 2019). Elsewhere in northern Thailand, as one tourist stated: 'Our first stop is Tha Phae gate, where one takes the Internet famous pigeon photograph for marking one's physical location on social media' (quoted in Gao et al., 2022, p. 7). Tourists' own images and conceptualisations of places were crucial. A focus on particularly desirable places, created though social media, helped to animate the tourist gaze (Urry, 1990) as was evident in posts from the Quad. Here we indicate how that extended from the University of Sydney to other places.

Diffusing discoveries and pleasures, and following others to interesting sites, is central to photography and to posting blogs and photographs on social media. Some of the instigators of this process were Chinese WHMs, with a little more time, motivated by photography and seeking to distinguish themselves and gain status by circulating novel photographs (Chap. 4). As one WHM stated, 'Last year I saw some beautiful pictures that my friends posted on WeChat and they told me they were doing a working holiday in Australia. I felt it so interesting, so I came to Australia' (quoted in Fang et al., 2021, p. 11). The pictures and experiences were catalysts for travel within Australia while their rationales were much like those of short-term tourists. But for some at least, seemingly a rare example of dissent from the norm, as one WHM pointed out: 'It's a pressure that I have to try these bucket-list activities in Australia. If not my friends would think it's a real pity' (ibid). No-one wanted to lose face by not doing what was expected. Otherwise, photography and tourism were inseparable. Moreover, photographs and posts, where they come from fellow Chinese millennials, were accepted as trustworthy, accurate and credible, and more authentic, reliable and unbiased than official publicity (Garner, 2020; Xiang & Gretzel, 2010; Zhou et al., 2023). Where it was feasible, following and replicating leaders made sense.

Gender and Influence

Although tourists who posted blogs and photographs had multiple reasons for doing so, from altruism to narcissism (see below), all effectively functioned as influencers (some deliberately so) at least in a small way. Influencers have become central to the *wanghong* (网红) economy of social media marketing, where *wanghong* (literally 'internet fame' or 'internet celebrity') refers to 'micro' celebrities and social media influencers, almost exclusively women, and often linked to fashion and beauty products (Abidin, 2018; Guan, 2021; Sandel & Wang, 2022). Thus social media and the Internet have contributed to the rise of 'micro-celebrities', famous within a small group, linked to self-commodification, but primarily in a context where some form of marketing is involved (Marwick & boyd, 2011; Guo, 2022). Some such influencers are thus professional, able to make a living from online promotion. Others are on the fringes of success, 'wild growing' (*yesheng*) without contracts with multichannel network companies (Guo, 2022). Bloggers and influencers of all kinds tend to be trusted simply because they are 'people like me'; and they may well be more trusted in the tourism contexts we discuss below where marketing is not involved and influencers are amateurs. Few of those posting blogs of travel destinations are much interested in commerce and the number of 'likes' suggests that 'celebrity' has rarely been achieved. While we cannot be sure, it is most unlikely that any of those who posted travel and photography observations made this even part of a career of influence.

Most of the posts analysed here, unless noted otherwise, were on Little Red Book, where they are less likely to go viral since the groups of recipients are limited. Some 91% of users of Little Red Book are younger women, and since the best estimates of independent travellers to Australia indicate that 62% are women (Austrade, 2019), it is reasonable to assume that most of the tourists in Australia posting there are women, and almost all the pictures are of women. Observations at various sites, and especially at the University of Sydney, confirm this assumption. Females were also more experienced travellers, travelling abroad more often than their male counterparts did, while also coming from more affluent families (Long et al., 2023). Men were relatively unlikely to be photographers, a situation also true within China (Liu, 2021, 2022). An analytical constraint is that it was difficult to ascertain in the next phase who were the authors of the various blogs and posts, although many left clues in their texts (and the preponderance of selfies assisted), but few provided motivations for posting their accounts and photographs. As photography has become more popular, it is a little easier to discern the sources of the photographs. What is often more difficult to determine, is how female tourists reached some obscure destinations in the first place.

Many younger women—millennials—engage in *guimi* holidays (闺蜜游) where two women, not in a relationship, travel together and share some interest in photography and fashion (e.g. Chen et al., 2021; Hao et al., 2021). Such holidays are 'characterised by the extensive use of and reliance on the internet, as some of them prefer recording their travel itineraries and sharing their emotions, memorable moments, and experiences online using, photos, videos and texts' (Chen et al., 2021, p. 232). Motivations for *guimi* tourism vary from the pleasures of independence, the opportunity for shared bonding, personal 'authentic' time, individualism and escapism. Being in a different realm, and the role of photography, was summarised by one *guimi* tourist:

> If I travel with my *guimi*, I took more photos and the photos are very nice. My boyfriend does not like taking photos at all and he always took me ugly when he took pictures. And sometimes he just could not understand why girls like taking so many photos (quoted in Chen et al., 2021, p. 235)

Visual observation and evidence from Sydney suggests that while not all *guimi* tourists were photographers, almost all photographers were *guimi* tourists, and many photographs were taken, checked and retaken at each site to achieve an excellent outcome. Through *guimi* tourism, and through photography, young women counteracted sociocultural constraints and gained freedom, prestige, pleasure and empowerment, as a form of feminist practice (Kong et al., 2022). That empowerment, evident in blogs and photographs, enabled them to play a part in influencing the travels and lives of others.

Beyond the obvious linkage between tourism and photography, where travel is documented visually, multiple reasons existed for why tourists then went on to post their photographs online. Various suggested factors range from personal development and self-enhancement, gaining prestige and status, social influence, social identity, and the sense of belonging to a particular group—evident through the 'likes' that follow posts (Qu & Lee, 2011) but, as often, it was simply the fun of recording their travel experiences and photographs (Ni, 2020), and the consequent pleasure of having followers, alongside the satisfaction of story-telling and blogging, and the tactile enjoyment of photography itself. Photography constructs an imagined authenticity that can be distributed into and through the 'statusphere', strengthening middle-class identities, creating social capital, and revealing social sophistication (Li et al., 2019; Kimber et al., 2019; Yin et al., 2020; Wu & Pearce, 2016). More such factors simply extended from those that earlier influenced bloggers, whose role as amateur influencers could be seen as their being part of a 'creative tourist class' (Gretzel & Jamal, 2009; Yoo & Gretzel, 2011). Some did see themselves as 'micro-celebrities'; others felt that they were providing a 'service' to other travellers and tourists and, having themselves benefited, were in turn posting in a spirit of ethics and reciprocity (Yin et al., 2020). Some rationales may have occasionally spilled over into narcissism and overly deliberate self-promotion especially where selfies are edited and photoshopped for the best presentation of self (Taylor, 2020; Tan & Yang, 2021). Ultimately some simply followed; as one Chinese tourist in Guangzhou stated, 'When I see my friends or the celebrities take photos in a nice place, I will visit that place and take similar photos' (quoted in Liu, 2021, p. 246). Incidentally, she might in turn have acquired her own followers.

Over time in Australia the presentation of self became more evident in the choice of fashionable outfits for the appropriate 'tourist pose'. Generally this matched a wider context where people felt pressured to compete and compare material possessions, sharing their experiences to impress others, but in a context where multiple values, both 'traditional' and 'modern', were simultaneously at play (Hsu & Huang, 2016). Social media are the obvious context for enhancing self-image, not least coming from places of particular aesthetic distinctiveness enhanced by dressing the part, most dramatically in wedding dress or Hanfu costume.

In China distinct urban and rural tourist sites are set up at popular tourist stops and decorated with background symbols and objects, ranging from imported alpacas and old cars to hanging bubble-chairs and mirror-covered platforms, where women might dress and pose for photographs, in colours, costumes and fashions both traditional and modern, but most recently often in Hanfu style. Much like *daka* tourism, this has also been a phenomenon of the late 2010s adopted by female millennials with young women dressing in Hanfu style (the traditional dress of the majority Han population of China, notably long robes) for photography. Wearing Hanfu clothing is almost absolutely dominated by women, for multiple reasons including entertainment, pleasure, aesthetics, culture, escapism, seeking cultural identity, a 'performance experience' and cosplay (costume play). Hair and make-up can be done on site, with the clothes also purchased or hired at the relevant sites, local 'Hanfu experience centres', to match the scenery (Shi, 2021). Hanfu wearing has become a popular trend for tourists in China and its practice has gone beyond mainland China, to include Taiwan, Singapore and Malaysia (Zong & Tsaur, 2024). The practice of dressing up is thus widespread and part of the rationale for tourism. The predilection for Hanfu, yet to reach Australia, demonstrated the widespread penchant for millennial women dressing up while being tourists, a context that has been transferred to Australia, but in more modern dress.

World of Weddings

In some respects, in Australia, the notion of dressing up and posing with distinguished backgrounds, emanated from wedding photography, and from the University of Sydney. In the 2010s, the University had become a stopping point on some group tours, attractive for its heritage architecture combined with an international reputation for tertiary education, and providing opportunities for photographs with old buildings, jacaranda trees and manicured lawns. It was not otherwise a site for 'standard' tourism and was entirely absent from guidebooks on Sydney (Chap. 5). That perspective changed in 2015 after the University became a popular location for photography. Two prominent Chinese people, both CEOs of major companies, had married in Cairns but posed for their official wedding photos in the Quad: the neo-Gothic core of the University (Fig. 6.1).

Fig. 6.1 Sister Milk Tea and Qiangdong Liu at the University of Sydney

Qiangdong Liu was the CEO of the internet e-commerce company, Jingdong, and his wife, Zhang Zetian, was a businesswoman, investor and chief fashion adviser of the luxury business of Jingdong. She was regarded as China's youngest female billionaire and had gained initial fame from the popularity of a photograph of her holding milk tea—to be nicknamed

Sister Milk Tea (奶茶妹妹), as the photograph became an internet sensation in China. Their fame and wedding photos were big news in China and made the University an immediate destination for wedding photography.

Tourists sought to have their own photos (often long post-wedding) taken there. Some couples had wedding photos taken in the Quad, while others dressed relatively formally for photographs, replicating Sister Milk Tea to encompass the imagined authenticity of the setting. We occasionally saw brides in corners of the Quad discreetly changing into wedding dresses—on top of their usual wear—brought from China in the aftermath of the wedding just for such photographs.

> I was amazed by the beauty of this building. I also met a few people coming to take wedding photos. Soooo romantic. Very beautiful sunshine in Sydney (16 January 2018)

Much like Sister Milk Tea, other popular Chinese stars chose other exotic overseas destinations, and their preferences for 'destination weddings' resulted in a preference for similar destinations as ordinary Chinese sought to follow suit. In the 2010s this emerged as a phenomenon where wedding photographs were often taken overseas at particularly beautiful sites, such as the Greek island of Santorini, constituting a distinctive form of tourist photography (Zhuang & Everett, 2018).

Within Australia, this phenomenon has contributed to substantially increased Chinese visitor numbers in the 2010s to a few distinctively photogenic sites, including Bridestowe Lavender Farm in Tasmania (Fig. 4.1) and Hutt Lagoon in Western Australia for wedding photographs, and more general photography. Just as Chinese wedding photography transformed the University of Sydney into a distinctive tourist site, so photography and the quest for colour created and transformed other sites (see below). Thus the Hutt Lagoon became the Pink Lake, where the CEO of the Western Australia Mid-West Development Corporation observed:

> Chinese brides are getting married in mainland China and often packing their wedding dress in a suitcase and catching a flight to Perth so they can come up to Pink Lake and wade into the water to take their wedding photos there (quoted in Guilliatt, 2020, p. 23)

At the University of Sydney, neither the wedding photos of Sister Milk Tea nor the magical vestiges of Harry Potter were known to a wider public, even as they became significant to Chinese visitors. As that occurred, the University was established as a 'marked place' and the overall perspective changed. For many Chinese visitors to the University, taking photographs became the main reason for visiting. 'Many people were taking photos when we arrived ... both inside and outside the Harry Potter building. We played with all kinds of poses for more than an hour' (4 January 2019, *Mafengwo*).

The photographic process that effectively began at the University of Sydney was not confined there. Moreover, most tourists are not on their honeymoons or necessarily with lifetime partners, especially the *guimi* tourists. Consequently, what are much more common are secular parallels to Chinese wedding photography, where destinations are not so far-flung, but sites are colourful and backdrops for more standard photographs: a little more prosaic, but more original, than Santorini. Photographers and their friends may still dress up but not in wedding dress. Nonetheless, as we outline below, sites were chosen as destinations specifically for their photogenic quality as backgrounds (rather than necessarily for some intrinsic 'value' of the site itself in terms of beauty, heritage value or historical significance).

Creating Alternative Tourist Sites

The University of Sydney became the first *daka* site in Australia but, in the second half of the 2010s, more such sites emerged, in or near the largest cities, yet significantly, in view of tourism practices hitherto confined to metropolitan regions, these sites, like Bridestowe and the Pink Lake, were far beyond the large cities.

However, despite Bridestowe, most of the emerging sites were within the metropolitan area of Sydney (and not at all obvious in Melbourne) and were not far from the central city. Indeed the earliest Chinese generated sites were reflections on (almost literally in the case of Lavender Bay, vis-à-vis the Opera House) and versions of existing tourist imaginaries of Sydney that generally centred on the inner-city area and with harbour and coastal attractions to the east of the city centre being prominent. Such iconic sites as the Opera House, Harbour Bridge and Bondi Beach, remained key destinations for Chinese tourists, but, for millennials at least, a group of 'marked places' augmented these icons. Based on the number of social media posts the more prominent of these can be identified (Fig. 6.2). However, crucially, for Chinese independent tourists they are

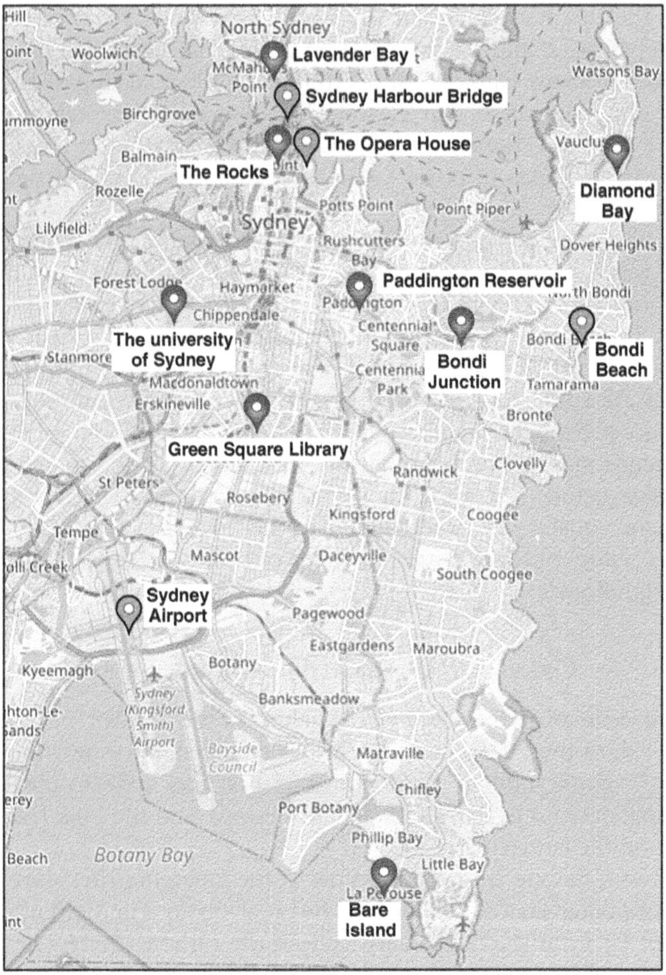

Fig. 6.2 Sydney. Iconic and *Daka* Tourist Sites

understood, experienced and communicated in ways that vary from traditional tourism, mainly because of photography and its connection with social networking platforms, and the desire to see these sites in different ways. Recurring themes among Chinese tourists posting to sites such as *xiaohongshu* were the opportunities to take distinctive photographs, avoid the crowds at iconic sites and exhibit a sense of discovery. Some of the reasons for discovering and enjoying new marked places are exemplified in one post on Lavender Bay, a site discussed further below:

> Every time visiting the Opera House, there are so many people that I can hardly take a single photo that does not look like a normal tourist photograph. This time I found a new location—Lavender Bay. It is not far from the Opera House, but has an excellent angle for photographing. Also there are far less people here. Very easy to capture amazing pictures (11 October 2019)

The new focus went well beyond the intrinsic merit of the sites, placing them in a harmonious photographic context.

We can now elaborate on this process of finding and photographing with particular reference to three recently established marked places in Sydney: Lavender Bay, Diamond Bay and Bare Island (Fig. 6.2). Like the University of Sydney, none of these places were mentioned in standard guidebooks of Sydney and there was almost nothing commercial at or near these sites. Observation indicated that, beyond Chinese tourists, other visitors were infrequent. Invariably the posts by Chinese tourists were positive and focused on aesthetic landscapes, almost always including water, and heritage. Broadly the same pattern recurred at most marked places: discovery, spreading the message of attractiveness to potential followers and making recommendations on social media, related to access, timing and colour, and often to safety. Social media empowered tourists, who collectively strongly influenced and even played a decisive role in the destination image.

Lavender Bay

Lavender Bay, facing the Opera House from the less visited northern side of Sydney Harbour, was frequently singled out as less crowded than the Opera House and a good site for photographs. An additional attraction was Wendy's Secret Garden, developed by Wendy Whiteley after her famous artist husband, Brett Whiteley, died in 1992.

> You won't miss taking a photo with the harbour bridge when coming to Sydney. If you want to avoid the crowds in Circular Quay also taking typical tourist photos, coming to Lavender Bay will make things different. You can

> get a full and good view of the harbour bridge, and enjoy taking beautiful pictures. Especially in the afternoon, the sunshine makes the bay shining, which gives a very different atmosphere. After taking pictures, you can stop at Wendy's secret garden to see different plants and relax yourself (15 March 2019)

That post also shared tips for transport and photography, and elicited responses:

> Looks fantastic. Thanks for sharing!
> Can I find it by searching Lavender Bay in Google map?

Another post took a similar perspective:

> I recommend this place where you can get a better view of the harbour bridge—Lavender Bay. It is getting more popular and becoming a 'marked place' because of the nice scenery of Sydney Harbour. It is also close to the Luna Park. Another tip is to have a visit to Wendy's secret garden. This secret garden is full of artistic temperament which will offer you an eye-opener. (28 March 2019)

That too brought comments: 'I visited there today, so much liked it; 'Looks so beautiful! Will go and mark'. The influence of Little Red Book is apparent in another post:

> The best location for photographing the Harbour Bridge. Never to be in the crowds again. I saw the recommendation of this spot on the RedBook, and I am amazed how she could have found this fantastic place. Most people go to Circular Quay to take photos. However, it is very crowded and the photos do not look very good. Lavender Bay is a small bay in North Sydney, surrounded by local residents. So there are many fewer tourists (10 February 2019)

Such a focus on peacefulness, harmony between land and water, natural and human-constructed landscapes and the central role of photography was replicated over and over again at Lavender Bay and at other *daka* sites (Fig. 6.3).

Fig. 6.3 Lavender Bay

Lavender Bay, is a beautiful niche daka spot. Lavender Bay is right next to Luna Park, about 5 minutes walk. This place is not crowded so you can take good pictures. You can see the entire Harbour Bridge, and it looks really beautiful

Being relatively accessible, Lavender Bay probably attracted more independent tourists than other *daka* sites (after the University of Sydney) and there was unanimity around what made it so attractive. As another visitor wrote:

> Lavender Bay, daka spot, picturesque scenery. Lavender Bay in North Sydney, I could never get tired of visiting it. The scenery here is like a painting. Every shot looks pretty. You should not miss coming here. Walk along the jetty: you can see the Harbour Bridge and Luna Park. Sailing boats slowly float in the bay, a very romantic harbour view. Here is not very crowded, so you can enjoy the scenery and peace. Simply Google Lavender Bay. It is located between Milsons Point and McMahons Point (10 August 2020)

Some balance of nature, water and built environment, and a backdrop of Sydney icons made it an irresistible site for visiting and for the creation of selfies.

Diamond Bay

A second site that had become popular by 2018 was Diamond Bay, a clifftop site north of Bondi with a stone archway ideal for photography and which resembled some of the constructed photography sites at Chinese domestic destinations. There were over one hundred posts in Little Red Book in a two-month period to late December 2019. The earliest known post, in April 2018, advised people that 'at Diamond Bay, there is a lookout, from where you can see the spectacular cliff and waves. Though I felt a bit scared, the view is totally worth it. Just keep in mind to be careful!'. The post continued by advising prospective tourists that it was 'easy to drive there or take a taxi (Uber). Or a cheaper way is to take the train or bus. When you get there, need to be patient to find the small way to the stone gate'. This post was influential because it was saved by many people, and visitor numbers increased rapidly along with posts on Chinese language social media platforms.

> Taking pictures on the cliff is soooo cool, but be careful! It is easy to get here following Google map. Free parking here. There is also a small stone gate where you can take very nice pictures. (16 October 2018)

> Coming to visit Sydney, if you want to avoid the crowds in those mostly visited tourism spots (like Opera House, Darling Harbour, Bondi beach), try the Diamond Bay. It is a very peaceful place. I visited on Saturday and I saw less than 10 people in 2 hours. The scenery is spectacular, with cliffs and rolling waves. Sometimes you can see people climbing on the cliff. It is a very good place for taking pictures. Taking red or white dresses will make you look nicer in the picture. (24 July 2018)

> This is a fantastic place, beautiful and quiet ... I especially like the waves hitting on the cliff. I saw many people coming to take pictures on the cliff and on a stone gate. I also highly recommend this spot. One thing to keep mind is the safety. Diamond Bay is not far from Bondi beach, so you can easily find it following the map. (5 November 2018).

A year later, a post on Diamond Bay lamented the tourism management that had subsequently been undertaken: 'Diamond Bay is so suitable for photographs. It is such a pity that now the rock stack has been isolated to avoid taking pictures in case of danger' (6 December 2019). Safety was an issue, with two women (one a Chinese tourist) killed in separate incidents at the site in 2019 and 2020, both falling over the cliff, a phenomenon that is no longer globally unusual (Weiler et al., 2021). Although the earliest known post in April 2018 warned about the dangers of the site, this did not prevent it becoming popular thanks to the influence of Redbook and photography (Fig. 6.4):

> Diamond Bay gets popular thanks to the RedBook. It is a very beautiful place, best for photographing (28 November 2018).

> Diamond Bay was only known to a small group of people early on. But after getting popular on the RedBook, so many people come to mark this place ... Be careful when taking pictures (2 July 2019).

Unlike some other ***daka*** sites, the spontaneous and unexpected emergence of Diamond Bay as a tourist destination became problematic since, as elsewhere, local authorities were caught unawares by the sudden increase in tourist numbers. No government body had managed the site for high levels of tourism. Posts from tourists warned of the dangers but were inadequate to prevent disasters. Management followed later.

Fig. 6.4 Diamond Bay

THE most beautiful bay in Sydney—Diamond Bay. Not even exaggerating to say it is THE most beautiful bay in the southern hemisphere. It is more like a private niche spot if compared to Bondi Beach. It is in fact not far from Bondi. Along the coastline from south to north are Bondi → Diamond Bay → Rose Bay → Watson Bay. (Hard to rank them, but I love Diamond Bay the most). Super spectacular won't be crowded even on the weekend. Take Bus 324/325 from Town Hall. Hope you all have fun

Bare Island

A third site that became an early marked place was Bare Island at La Perouse, to the south of the Sydney city centre, a rocky outcrop on the north shore of Botany Bay (and linked to the coast by a wooden pedestrian bridge). Despite its extensive Indigenous history, and being associated with colonial exploration and settlement (evident from the French name La Perouse), it became popular via Little Red Book, with around forty posts in December 2019, primarily because it is one of the few places in Sydney—and the only one not far from the city centre—where it is possible to see sunsets over water. It therefore became recommended for photos, beginning with the earliest known post in May 2018:

> The most beautiful sunset in Sydney. Remember to check the sunset time before heading there... Very very beautiful sunset. The photos do not require any polish (2 May 2018)

> This is the marked place of the most beautiful sunset in Sydney. It is the filming location of Mission Impossible. It has witnessed countless couples taking wedding photos. Easy to get here with both public transport and self-driving ... The lighthouse, the unique rocks, the long wooden bridge, the rolling waves all look good in photos. Should not miss this place (25 March 2019).

> Niche marked place in Sydney, La Perouse. This is probably the most beautiful and coolest place I've ever visited over this month in Australia. La Perouse memorialises the first landing of French people in Australia. It is the filming location of Mission Impossible. A very good place to relax. It is also said to be the marked place of the most beautiful sunset (15 October 2019).

Bare Island was also popular as a destination for wedding photographs:

> When taking wedding pictures in Sydney, one place you must consider is the La Perouse beach. It is the filming location of Mission Impossible. The rocks look very remarkable. There is a bridge, beneath which the scene is also very nice (28 November 2019).

Here too, like the other *daka* sites, Bare Island was popular because of its photographic potential and the diversity offered by an island and wooden bridge, tidal changes and the colours of sunset and, atypically, with a link to *Mission Impossible*. Of all the *daka* sites it was the most popular for wedding photography (Fig. 6.5)

Fig. 6.5 Bare Island

Amazing! Sydney Daka spots are great for photos!!! La Perouse Beach and the Opera House are the most popular spots in Sydney for shooting wedding photos. I personally think La Perouse can get lots of great photos! If you haven't been, please go!!!

Other *Daka* Sites in Australia

While the previous three sites were relatively popular and attracted many *guimi* and other independent tourists, they were not the only ***daka*** sites in Sydney. Other marked places in Sydney included Paddington Reservoir Gardens, the Green Square library and the Pink Wall in the Rocks (Fig. 6.2). They shared many of the characteristics of other ***daka*** sites: colour, accessibility, the lack of crowds, and the space and place to engage in photography and some degree of performativity.

Paddington Reservoir Gardens, off Oxford Street, once a source of water to the city, was reconstructed in 2009 by salvaging original brick, timber and iron fixtures to make a sunken garden, with graffiti art preserved in its eastern chamber. Despite Sydney City Council's description of it as being 'Hailed as a blend of the ancient Baths of Caracalla in Rome and the Hanging Gardens of Babylon', and recommending that it be visited at sunset it never became a standard tourist destination. Nonetheless, a decade later, posts referred to it enthusiastically: Paddington Gardens has 'a super beautiful garden, perfect for vintage style and aesthetic photos' (23 January 2020) where it is 'better to dress in dark clothes' (8 July 2018).

Two other sites within Sydney that belatedly became ***daka*** sites were the Green Square library and the Pink Wall in the Rocks. The modernistic Green Square library, opened in 2018, proved an attraction because of its 'rainbow wall' bookshelf, where books were placed according to the colour of the spine rather than by subject.

> This library is the newly declared marked place in Sydney. Super beautiful rainbow bookshelf. Perfect for both study and photographs (11 April 2019)

> I could not imagine that there is an underground library with a very fascinating design near the city centre. Especially there is a rainbow wall which attracts many people to take photos (14 September 2019)

The only mildly negative post on any ***daka*** site related to the library ('The rainbow bookshelf looks extremely beautiful on photos. But I felt it was not fully worthwhile coming here only for this small bookshelf' (30 September 2019)). A line of pink painted houses (the Pink Wall) in the Rocks, the oldest heritage area in Sydney, became popular too.

> It has become popular because some professional photographers found it and used it as a wedding photograph location. Gradually it became more and more popular and became a marked place … It is not in bright pink so you probably need some post-processing to make the photos look nice (21 February 2020)

Just as single walls became significant at Green Square and the Rocks, so a single staircase in the Queen Victoria Building (a remodelled nineteenth century heritage-listed shopping centre, in the heart of the city) became a destination, because of the juxtaposition of the elegant winding wooden staircase and the sun refracted onto it through adjoining stained-glass windows.

Some tourist sites favoured by Chinese visitors were ephemeral according to season, largely because seasonality influenced colour. McDougall Street in North Sydney acquired seasonal popularity as a *daka* site from 2017 through its concentration of blooming jacaranda trees whose mauve flowers inspired photography. At the University of Sydney, jacaranda trees on the main campus were also a popular backdrop for photographs in late Spring and early Summer. By contrast, the iconic Blue Mountains, outside Sydney, were most popular in Autumn, as the leaves change colour. The golden leaves of introduced deciduous vegetation, combined with native eucalyptus trees that create a blue haze (which inspired early settler-colonial explorers to name the mountains), offers many opportunities for photographs.

Not all suggestions were taken up. Hornby Lighthouse, though accessible near the South Head entrance to Sydney Harbour, seemingly failed to attract followers despite the opportunities:

> A super great place for photography. With the lighthouse plus the sunset, you can take fabulous photos so easily. Not only suitable for friends but also for couples to take wedding pictures. For dress style, long dresses and Bohemian style dresses are good choices. Better to arrive before 5pm to get the best sunlight. Other than this lighthouse, there are some places nearby also very suitable for photographs (19 May 2019)

The lighthouse was however included on another post under the heading 'A list of niche tourism spots. Come before they get popular' which also included Bombo Headland ('Walking though the basalt pillars makes me feel like walking on the moon. Takes about two hours by car') and Cape Solander ('200 metre cliffs make this the best lookout. It is also one of the best places for watching whales') (22 June 2019). Bombo was simply too far away but Cape Solander (Cronulla) did belatedly become a *daka* site. Moreover, somewhat as had occurred at the University of Sydney, it also subsequently attracted other tourists after Chinese tourists had paved the way, emphasised grandiosely in the Australian version of the *Daily Mail* ('Why adventure-seekers can't stop talking about this breathtaking 'hidden cave' just 45 minutes from a major Aussie city', 22 February 2023) which noted that it was 'off the beaten trail … and somewhat difficult to find' and that 'travellers should not attempt to reach the spot during days of high sea level': directions analogous to those on Red Book. The key feature of the site was a 'hidden cave' and doorway with a circular design.

Marked places also exist in Greater Sydney, including Long Jetty, on the Central Coast north of Sydney, and Helensburgh to the south of the city, with the former popular for its eponymous jetty and the latter for its glow worm caves. At Long Jetty

> It is a nice place to see both sunrise and sunset. It is also a place you must go to when visiting Sydney. It is a bit far from the city, and that is why there are not many visitors around. The sea is very quiet and peaceful. A nice place to visit (12 November 2019)

By contrast, Little Purry Jetty in Wollongong was recommended, but, like Bombo, was too distant, not easily accessible, and not distinctive enough. So too was Anna Bay (Port Stephens) despite 'during the sunset, the sky is dyed to a rosy red, which looks glamorous and is very suitable for photographs' (2 December 2019). Typically at all these sites, there were no restrictions on the time people could stay there, and they were scenic, safe and ideal for photography and—with the exception of Long Jetty—easily and quite cheaply accessible.

Similar *daka* sites may also exist in other large Australian cities, with colour and water being important attractions. In Perth, the Crawley Edge Blue Boatshed, jutting into the Swan River, was easily accessible yet, like many other *daka* sites, its sudden significance was thoroughly surprising to local residents, evident in the title of one account: 'How a humble Perth boathouse became Australia's most unlikely tourist attraction'

(Wilk, 2019). The blue boatshed was popular for its colour and for how changes to the water, sky, clouds and light offered many photographic opportunities. It had been derelict for many years and after restoration was ignored by residents and other tourists. Remarkably, the 'global destination brand awareness' that discovery by Chinese tourists had created was 'just priceless' Leaver and De Poloni (2019) with the site later used by Singapore Airlines, JAL and Qantas in Japanese advertisements, so boosting the local economy.

Most surprisingly, given the reluctance and inability of most tourists to travel beyond the metropolitan fringes, some popular destinations emerged in very remote areas in states that most Chinese tourists did not visit (and in places that had attracted few other tourists, international or domestic). Foremost amongst these has been Lake Tyrrell, a remote, shallow salt lake near the small town of Sea Lake in thinly populated north-west Victoria, a considerable distance (about 400 kms) from Melbourne. In 2014 a local resident had sent a photograph of two people silhouetted against sunset over the lake to an agent in Sydney who specialised in Chinese tourism. The picture was loaded onto Instagram and quickly went viral on WeChat. Within three days, tourists started arriving and a minor economic revival followed with cafes opening and new accommodation being constructed. The lake was particularly distinctive at dawn and dusk when its salt crystals and crust, and shallow water, turned it into a reflective mirror where there was no obvious boundary between earth, lake and sky (Fig. 6.6) and clear night skies enabled photographs to be composed that included the stars (Grindlay, 2016; Wright, 2019).

Equally prominent, at least in media coverage, was the Pink Lake (Hutt Lagoon) in Western Australia, despite its being 500 kms north of Perth (Fig. 6.7). Like the Blue Boathouse and Lake Tyrrell, it initially became popular on Instagram around 2014 and was used in cinematography, music videos and television commercials. Its presence in a Lancôme lipstick advert, with a model wearing a red satin ballgown, was an initial catalyst. It was visited in 2019 by Meng Fei, also known as 'Grandpa Meng', the host of the popular Chinese prime time television dating show *Fei Cheng Wu Rao* (*If You are the One*). One blog advised visitors to come on a sunny day between 10 am and 2 pm to get the most vivid pinkness, to be aware that the water levels are lower in summer and the colour is further away from the shore and, if wading in, to wear 'reef booties' to protect feet from the sharp salt. The pilot of a local plane company, whose

Fig. 6.6 Lake Tyrrell, Victoria. Photo by Luke David

Fig. 6.7 Pink Lake, Western Australia *(ABC Midwest and Wheatbelt: Cecile O'Connor)*

business had boomed, observed simplistically: 'The pink colour is romantic to them and of course they love the clear skies and the clean, white clouds' (ABC, 2019). Other destinations, such as Uluru, Coober Pedy and Mount Gambier, have had some limited success with Chinese tourism, but were not *daka* sites that Chinese tourists themselves had pioneered.

Colourful places have dramatically boosted Chinese tourism in a few parts of urban and rural Australia, and in the most improbable sites and destinations, to the extent that whether in inner cities or remote rural areas a localised and sudden over-tourism has occurred with local authorities having belatedly to provide signage, safety and toilet facilities. Unlike the Quad at the University of Sydney, rigorously curated (but not for tourism), none of the sites were managed as tourist destinations. At least in the rural destinations, especially Sea Lake and Kalbarri (close to the Pink Lake), tourism boosted stagnating economies. Nonetheless, most *daka* sites are in or close to the major cities, although particularly colourful sites such as the Pink Lake (and Bridestowe Lavender Farm, a forerunner) were exceptions to the rule. The drama of colour in culture and photography was crucial in motivating Chinese tourists to sometimes travel many hours to see, photograph and be photographed in front of a tourist attraction whose appeal was derived primarily from its colour. Whether such places prove to be ephemeral as trends and tourists change remains to be seen.

Planning for Colour

Since Little Red Book's audience (and 'market') is primarily young, female and 'high-end' consumers, and those who post there are the same, it may be 'natural' to 'choose a 'fashionable' theme when producing and symbolizing travel pictures' (Zhou et al., 2023, p. 6). As one such photographer stated: 'Most followers on Red pursue the fashion culture and value the quality of life. So we usually select some romantic B&Bs or scenic sites. Even if we go to a traditional scenic area, we will also seek and highlight its stylish side' (quoted in Zhou et al., 2023, p. 6). Style, colour, content and context have thus been important in both creating the new tourist sites, oriented to fashionability, and ensuring the appropriate performativity at the sites.

Colour has long been recognised as a marketing tool, including for the Chinese tourism market in Queensland (Chap. 3), but has been evaluated almost exclusively in terms of its impact on choice of destination, and then linked to retail packaging and marketing (Yu et al., 2020). Destinations may be described as colourful, but actual colours are unimportant, yet colour influences tourist experiences and emotions (Yu & Egger, 2021). No literature discusses colour in terms of the dress style and performativity of the tourists themselves. That they themselves add colour to the destinations, and deliberately do so, has been ignored.

While many tourists prefer to be photographed in adequate, appropriate, smart and even elegant clothes, that has been taken to a higher level by Chinese *guimi* tourists, evident in similar practices within China (Liu, 2021) and in even more dramatic form in Hanfu tourism. It was readily evident that many female Chinese tourists had dressed in styles and colours that they deemed appropriate to the site they were visiting and the image they sought to portray. That was particularly so for wedding photos, which required a certain gravitas in a relatively formal setting (Fig. 6.5). Equally tourists chose, and were encouraged to choose, clothes and styles appropriate to the season or time of day. By contrast dramatic colourful lakes could be a background to almost any style and colour. The predilection for dressing up colourfully, whether in wedding garb, stylish modern dresses, or even in brilliant red Hanfu robes, was invaluable for matching the scenery.

Colour is a dynamic concept with cultural connotations and spatial and temporal dimensions, but with colours 'invested with different meanings in China than in Western cultures' (Kommonen, 2011, p. 367) despite a growing convergence. Vivid pink lakes, mauve jacaranda blossoms and the purple lavender of Bridestowe (parallels to domestic cherry blossom tourism) combine Chinese tourists' love of Australian nature, photography and performance in exceptionally colourful settings: prime examples of the importance of colour, especially to the independent *guimi* tourists. The importance of colour was frequently reiterated in Little Red Book posts: it was valuable to be directed and properly prepared to take photographs. Chinese tourists researched sites in advance, trusting the advice of their predecessors, and planned their photographs well before they actually reached the site. That was early evident at the University of Sydney in the wake of Sister Milk Tea's wedding photographs. Typically, one post advised:

> If you are aiming to take nice photos at the University of Sydney, known as one of the film locations of Harry Potter, better to avoid May to August as it is winter (contrary to the northern hemisphere). There are usually lots of

> rainy days in winter ... Suggestions for clothing: ... don't wear clothes with a mix of many colours. The Quadrangle Building is a more historical style building, and goes better with decorous colours rather than bright fancy colours (5 April 2019)

As one female tourist (25–29), who had previously been a student at a different university in Sydney, said: 'We wanted to look like students so we came in our tee-shirts, jeans and hoodies. Unfortunately we chose the wrong colours so we didn't match the front of the building'.

The wedding photos of Sister Milk Tea soon gave way to the imagined authenticity of the Quad as the site of the Harry Potter building as the main rationale for visiting and photographing. To celebrate their presence and the 'authenticity' of the site, some tourists had brought their own accessories, including gowns, wands, books (and other Potter ephemera), and 'performed' magic or sought 'to fly', to a Hogwarts background, to participate in the stories and further enhance the destination. Photography ensured their performances were captured in social media posts, as they enacted a more sensual engagement. The University of Sydney thus exemplified a *daka* site that should be visited, where colour, the right dress, the pose and performance were vital for photo taking.

By the end of the decade it was the leading *daka* destinations for millennials, with social media posts reflecting the merging of the Harry Potter myth, the heritage architecture and the landscape, with the scope for and practice of photography:

> The main purpose of coming to the University of Sydney is to see the legendary Harry Potter House. So grand. In every place I want to take pictures. There is a large lawn behind the Harry Potter House, and there is a feeling of calmness in the years. (19 November 2018)

> The University of Sydney has a beautiful environment, close to the city, convenient transportation, full of blue-flowered buds [jacaranda] everywhere, and the exam season is blooming. (28 February 2019)

> See the main goal of the University of Sydney trip—Harry Potter House ... After two days of rainy baptism, today Sydney is finally clear, blue sky and white clouds, setting off the brown retro building, but also the feeling of medieval movies. There is a large lawn in front of the main building, which

> is very suitable for taking pictures. We also saw a couple of wedding couples. Today, the day is beautiful, both inside and outside the Harry Potter House, even playing with the film, all kinds of poses for more than an hour. (4 January 2019)

Social media advice extended to recommending what times were best to take photographs, when the campus was peaceful and what was required in order to obtain the desired photographs without too many people intruding. Independent travellers were able to benefit from having their own itinerary and time.

> If you want to see a retro campus in Sydney, of course you must come to Sydney University. The historic building is famous as the Harry Potter building. It is also a popular place for wedding photography. 3pm – 4pm is the best time for photography (26 March 2019)

> It's definitely a must-have place in Sydney. We spent a lot of time at the University of Sydney, mainly looking for a suitable camera angle, but in the case of a large number of visitors, the only thing you can do to make the photo background 'cleaner' is to wait (29 January 2019)

Tourist posts usually commented on the aesthetic qualities of the site, such as open spaces and heritage architecture (and, of course, the resemblances to Hogwarts) and the green grass in the Quad and beyond, all of which made it an excellent site for photographs. Many suggested that it was important to complement the aesthetics of the site by visiting at the right time ('good for afternoon shots') and wearing the right colours and dresses, suggestions exclusively directed at women. Our observations suggest that many tourists complied. Some even recommended a Spring visit when the jacaranda trees in the Quad were in bloom, and advised relatives in China accordingly. That some of these suggestions needed forethought and planning, even before leaving China, notably bringing particular articles and colours of clothing, emphasised the significance of following, posing and performing in the appropriate manner. Similar directions and suggestions followed from other marked place, unsurprising since they had been chosen for their photogenic qualities. In a sense creativity was followed by compliance.

The Colours of Tourism

The rise of independent Chinese tourism in Australia amongst growing middle-class millennials, and especially young women, has resulted in the emergence of new distinctive tourist sites. That has been assisted by improved English language skills, the growth of longer-term student and resident migrant populations, but above all by smartphones. Pictures and videos posted on social media, especially *xiaohongshu*, have become significant in shaping—even creating—destination images, and inspiring subsequent visits. The identification of marked places, *daka* sites, unknown or rarely visited by other tourists highlights a distinctive, dynamic and gendered Chinese tourist geography of Sydney involving extensive use of social media, before, during and after tourism, and independent of, and different from, that portrayed in traditional tourist guidebooks.

Social media and Chinese tourism have combined to create a distinct and rapidly evolving tourist geography of Sydney. A few Chinese tourists have posted on Facebook sites in English; some posts have been read by Chinese speakers from countries such as Singapore and Malaysia, and that has attracted other tourists to some sites. Local residents too are familiar with most sites. In the middle of Sydney exclusivity could hardly be possible. Yet, at two remote lakes, the unexpected arrival of tourists boosted stagnant local economies. An earlier consensus that Chinese tourists prefer to visit major attractions (Liu et al., 2019), remains true, but more recent independent tourists, guided in various ways by their smartphones, have significantly contributed to creating new, quite different and more personal tourist spaces. Yet, by following the trail of newly marked sites—however selectively—many millennials are actually following a 'tour', albeit radically different from old fashioned whistle-stop group tours.

Smartphones have quickly gone far beyond a basic value, providing easy access to information on hotels and transport, to not merely recording tourist trajectories but significantly changing them: creating a rich, gendered alternative tourist geography of Sydney and Australia. So fast changing is this geography, as it enables new discoveries, that it has posed problems for site management (notably at Diamond Bay). That fluidity and flexibility is evident, with one tourist posting from an unusual, distanced perspective on the back of the Quad: 'No—don't stand at the Quad—follow me to other Sydney University sites' (20 September 2020). Another made a similar suggestion from a rather greater distance (Fig. 6.8). The Quad however may prove not to be ephemeral but simply a starting

Fig. 6.8 University of Sydney: Behind the Quad

The Harry Potter Building has been shot many times. Although it is pretty, the photos are all the same. I looked for where was best to take different photos on purpose and I took some today. Indeed these other places are good for photos. I am sharing with everyone here!

point in the quest for distinctiveness. As the rare negative comment on Green Square library suggested, some sites may however be ephemeral and others, like the Little Purry Jetty in Wollongong, simply too distant and inaccessible, and perhaps too much like other sites, to become marked places.

The significance of these new sites, albeit for a Chinese audience, demonstrates the power of tourist agency through photography to depict and share new destinations, in contrast to the role and authority of conventional media and marketing agencies. Since it is invariably the best photos that are posted, that increases the range and number of possible attractive sites in destinations. Beyond marketing of places, these aesthetic aspects of tourism have often been overlooked. Chinese millennials are responsive to the intensity and patterns of colour, and many go to great lengths in planning their visits to add their own colour to the destinations.

Photographing and performing at the Quad repeated and enhanced the myth of Harry Potter and reinforced soft heritage, while elsewhere photography created new places and spaces, that offered discovery and therefore authenticity. ***Xiaohongshu*** has, however, been criticised in China for being 'filled with false travel pictures that misled consumers by adding too many filters and coloring' and which could even be regarded as 'landscape alienation in virtual space' (Zhou et al., 2023, p. 1), but a careful review of posted pictures of the sites, all familiar to the authors, suggested that this was not evidently so. With the single expression of disappointment at Green Square Library (as much about quantity rather than quality) no subsequent visitors expressed any disappointment at any of the sites. Being a fellow woman millennial had advantages; influencers were 'like us'—posting and blogging for fun—perhaps making themselves micro-celebrities, but simply amateurs, and thus more trustworthy arbiters of harmony and authenticity.

Place is crucial to understanding independent Chinese tourism. Those that are scenic are particularly attractive, especially at daily or seasonal moments. Equally as important is the need for tourists to situate themselves in these places: where they 'feel unique in their travel experience, the more likely they will share those experiences in social media' (Ek Styvén & Foster, 2018, p. 88), as sharing reflects positive experiences and evaluations, often highly enthusiastic, and informative (how to reach sites, what to wear, when to go … how to be safe) and thus likely to encourage others to follow.

Travel photography goes beyond capturing the beauty of destinations (or the tourists) to shape destination images—and the annotations, guidance and directions enhance this—to inspire travelling and repetition. As visual content increases in popularity, aesthetics has become ever more important in conveying a certain lifestyle. The role of user-generated content and photography is likely to only increase further. It remains an open question whether these sites will, at some point, become part of a wider tourist trajectory, as eWOM and photography extend beyond their present concentration amongst Chinese millennials, or fads, fashions and tourist preferences will move onwards and elsewhere.

Bibliography

ABC. (2019, September 12). Selfies, wedding dresses and campers: China's big crush on Port Gregory's pink lake. *ABC News.*

Abidin, C. (2018). *Internet celebrity: Understanding fame online.* Emerald.

Anon. (2019, August 17). Daka destinations. *The Economist*, 30.

Austrade. (2019). *Chinese free and independent travellers – Their potential for regional Australia.* Austrade.

Chen, X., Mak, B., & Calder, I. (2021). The Guimi holidays of young Chinese women: Evidence from Cyber Communities. *Journal of China Tourism Research, 17*(2), 230–248.

Edensor, T. (1998). *Tourists at the Taj: Performance and meaning at a symbolic site.* Routledge.

Ek Styvén, M., & Foster, T. (2018). Who am I if you can't see me? The 'self' of young travellers as drivers of eWOM in social media. *Journal of Tourism Futures, 4*(1), 80–92.

Fang, Y., Chien, P., & Walters, G. (2021). Understanding the emerging Chinese Working Holiday market. *Journal of Vacation Marketing, 27*(1), 3–16.

Gao, J., Zeng, X., Zhang, C., & Porananond, P. (2022). Understanding the young middle-class Chinese outbound tourism consumption: A social practice perspective. *Tourism Management, 92*, 104555.

Garner, A. (2020). I came, I saw, I selfied. Travelling in the age of Instagram. In M. Mansson, A. Buchmann, C. Cassinger, & L. Eskilsson (Eds.), *The Routledge companion to media and tourism* (pp. 316–325). Routledge.

Gretzel, U., & Jamal, T. (2009). Conceptualising the creative tourist class: Technology, mobility and tourism experiences. *Tourism Analysis, 14*(4), 471–481.

Grindlay, D. (2016, March 24). Chinese tourists flood isolated grain town, throwing drought-stricken community an unexpected lifeline. *ABC Rural.*

Guan, Z. (2021). Chinese beauty bloggers: Amateurs, entrepreneurs, and platform labour. *Celebrity Studies, 12*(2), 326–332.

Guilliatt, R. (2020, January 25). Invasion of the insta tourists. *Weekend Australian Magazine*, 20–23.

Guo, J. (2022). The postfeminist entrepreneurial self and the platformisation of labour: A case study of yesheng female lifestyle bloggers on Xiaohongshu. *Global Media and China, 7*(3), 303–318.

Hao, F., Zhang, S., & Xiao, H. (2021). A return to innocence: Guimi tourism and women empowerment. *International Journal of Tourism Research, 23*(6), 971–983.

Hsu, C., & Huang, S. (2016). Reconfiguring Chinese cultural values and their tourism implications. *Tourism Management, 54*, 230–242.

Hwang, Y. (2011). A theory of unplanned travel decisions: Implications for modeling on-the-go travelers. *Information Technology and Tourism, 12*, 283–296.

Kimber, S., Yang, J., & Cohen, S. (2019). Performing love, prosperity and Chinese hipsterism: Young independent travellers in Pai, Thailand. *Tourist Studies, 19*(2), 164–191.

Kommonen, K. (2011). Narratives on Chinese colour culture in business contexts: The Yin Yang Wu Xing of Chinese values. *Cross Cultural Management, 18*(3), 366–383.

Kong, S., Guo, J., & Huang, D. (2022). The girlfriend getaway as an intimacy. *Annals of Tourism Research, 92*, 103337.

Larsen, J. (2005). Performativity of tourist photography. *Space and Culture, 8*(4), 416–434.

Leaver, K., & De Poloni, G. (2019, June 30). How a little blue boathouse in Perth became the hottest selfie spot on Instagram. *ABC News*. Accessed August 10, 2023, from https://www.abc.net.au/news/2019-06-30/perth-blue-boathouse-in-crawley-becomes-instagram-star/11261312

Li, M., Sharpley, R., & Gammon, S. (2019). Towards an understanding of Chinese tourist photography: Evidence from the UK. *Current Issues in Tourism, 22*(5), 505–521.

Lin, Z., & Yang, L. (2021). Smartphones as actors: A new digital disability care actor-network in China. *International Journal of Cultural Studies, 24*(4), 673–688.

Liu, C. (2021). Exploring selfie practices and their geographies in the digital society. *Geographical Journal, 187*, 240–252.

Liu, C. (2022). Imag(in)ing place: Reframing photography practices and affective social media platforms. *Geoforum, 29*, 172–180.

Liu, Y., Huang, K., Bao, J., & Chen, K. (2019). Listen to the voices from home: An analysis of Chinese tourists' sentiments regarding Australian destinations. *Tourism Management, 71*, 337–347.

Liu, X., Wang, D., & Gretzel, U. (2022). On-site decision-making in smartphone-mediated contexts. *Tourism Management, 88*, 104424.

Long, F., Aziz, N., & Ngah, A. (2023). Exploring the relationship between travel motivations and information sharing behaviors on social media: Gendered differences of Chinese Gen Y in a context of Confucianism. *Journal of China Travel Research, 19*(2), 245–265.

Madionou, M. (2016). Ambient co-presence: Transnational family practices in polymedia environments. *Global Networks, 16*(2), 183–201.

Marwick, A., & boyd, d. (2011). I tweet honestly, I tweet passionately: Twitter users, context collapse and the imagined audience. *New Media and Society, 13*(1), 114–133.

Ni, J. (2020). *Daka travel experience: An examination of Chinese Millennial travellers.* MSc thesis, Purdue University.

Perkins, H., & Thorns, D. (2001). Gazing or performing? Reflections on Urry's tourist gaze in the context of contemporary experience in the antipodes. *International Sociology, 16*(2), 185–204.

Qu, H., & Lee, H. (2011). Travellers' social identification and membership behaviours in online travel community. *Tourism Management, 32*(6), 1262–1270.

Sandel, T., & Wang, Y. (2022). Selling intimacy online: The multi-modal discursive techniques of China's *wanghong*. *Discourse, Context and Media, 47*, 100606.

Shi, S. (2021). *Travelling with Hanfu: A social media analysis of contemporary Chinese travelling for artistic photographs.* MIHM thesis, Auckland University of Technology.

Stylianou-Lambert, T. (2012). Tourists with Cameras: Reproducing or Producing? *Annals of Tourism Research, 39*(4), 1817–1838.

Tan, W., & Yang, C. (2021). The relationship between narcissism and landmark check-in behaviour on social media. *Current Issues in Tourism, 24*(24), 3489–3507.

Taylor, D. (2020). Putting the 'self' in selfies: How narcissism, envy and self-promotion motivate sharing of travel photos through social media. *Journal of Travel and Tourism Marketing, 37*(1), 64–77.

Tussyadiah, I., & Fesenmaier, D. (2009). Mediating tourist experiences: Access to places via shared videos. *Annals of Tourism Research, 36*(1), 24–40.

Urry, J. (1990). *The Tourist Gaze.* Sage.

Weiler, B., Gstaettner, A., & Scherrer, P. (2021). Selfies to die for: A review of research on self-photography associated with injury/death in tourism and recreation. *Tourism Management Perspectives, 37*, 100778.

Wilk, V. (2019, June 21). How a humble Perth boathouse became Australia's most unlikely tourist attraction. *The Conversation.*

Wright, T. (2019, April 27). 'Ghost' town booms as hordes hit remote lake to capture magic moments. *The Age.*

Wu, M., & Pearce, P. L. (2016). Tourism blogging motivations: Why do Chinese tourists create little 'Lonely Planets'? *Journal of Travel Research, 55*(4), 537–549.

Xiang, Z., & Gretzel, U. (2010). Role of social media in online travel information search. *Tourism Management, 31*, 179–188.

Yin, J., Brown, G., & Huang, S. (2020). Social media and Chinese tourist behaviour. In S. Huang & G. Chen (Eds.), *Handbook on tourism and China* (pp. 91–105). Elgar.

Yoo, K., & Gretzel, U. (2011). Influence of personality on travel-related consumer-generated media creation. *Computers in Human Behavior, 27*(2), 609–621.

Yu, J., & Egger, R. (2021). Color and engagement in touristic Instagram pictures: A machine learning approach. *Annals of Tourism Research, 89*, 103204.

Yu, C., Xie, S., & Wen, J. (2020). Coloring the destination: The role of color psychology on Instagram. *Tourism Management, 80*, 104110.

Zhou, M., Lu, X., & Wang, B. (2023). Travel pictures authenticity in social media space. *Annals of Tourism Research, 101*, 103581.

Zhuang, Y., & Everett, A. (2018). A brief history of Chinese wedding and bridal photography tourism: Through the lens of top Chinese wedding photographers. In E. Yang & C. Khoo-Lattimore (Eds.), *Asian cultures and contemporary tourism* (pp. 79–100). Springer Nature.

Zong, Y., & Tsaur, S. (2024). Destination experience scape for Hanfu Tourism: An explanatory study, *Journal of China Tourism Research, 20*(1), 144–166.

CHAPTER 7

A Colourful Future?

Chinese tourism to Australia has reflected and embodied many of the trends of the twenty-first century—the rise of a more affluent middle class, technological progress (especially in photography and social media), the significance of both popular culture and geopolitics, new gendered developments and differences and the importance of transport routes. We summarise and review here key trends and themes, from mass tourism to the rapid evolution of a new cultural and geographical scene directed by social media and centred both on the greater individualism of 'following the heart', but also, seemingly paradoxically, involving marked places, recommendations and directions, and new itineraries. These trends have shifted the actuality of Chinese tourism in Australia beyond the wider literature on tourism, which takes little account of significant and distinctively Chinese cultural themes. We have approached this by looking at these fascinating changes, mainly within a single city, so enabling new understandings of the phenomena of Chinese tourism, centred on the practices and strategies of the tourists themselves.

Through developing an understanding of the recent history, context and impact of Chinese tourism, we have drawn out how relationships between two countries can be perceived through tourism, bleakly so in the years of COVID-19 (when trade between the two countries faltered). President Xi's vision and invocation of the Chinese Dream in 2013, where young people should dream, think and work to revitalise the Chinese

J. Connell et al., *Chinese Tourism in Australia*,
https://doi.org/10.1007/978-981-97-2477-2_7

nation, motivated policies and external orientation in areas such as education and tourism. The geopolitics of tourism resurfaced in the post-COVID era when other countries, including New Zealand, received Approved Destination Status for the revival of group tourism, but Australia was excluded. Independent millennials however returned as individuals, following their hearts and dreams.

While social media offer rich data on and from a rapidly changing phenomenon, they help to elucidate only one particularly prominent 'alternative' direction within a rapidly changing metropolitan tourism context. Our sample of tourists was relatively small (though typical of independent tourists) but provides a valuable perspective on how international tourists' social media use may contribute to and shape a new tourism geography. No formal sampling of users of eWOM was undertaken, and much more needs to be known for a greater understanding of tourist attitudes, the role of shared knowledge, values, relationships and their perceptions of value and significance. Some tourists are more interested, motivated and prepared to engage in advocacy than others are but we know little about which ones and why some prove to be key influencers, and why gender differences are so profound. Much the same is true of reception of eWOM messages. Yet even a small-scale qualitative study identifies and analyses remarkable externally inspired influences on metropolitan Sydney, where Chinese enthusiasm for Harry Potter has created a distinctive site: representing the power of a fictional character to reconstruct notions of authenticity and 'soft heritage', draw in international fans, and subsequently inspire and become the first component of a new tourism geography.

Since China gained ADS in 1999 for tourists from its three largest cities, Chinese tourism has evolved from stereotype packaged tours to become more complex, subtle and differentiated. At the start of the century Chinese tourism was tentative, conventional and typical of what had drawn other previous visitors—nature and iconic sites—and caught up in the tourist bubbles of package tours, often involving groups from the same workplace. It was much like tourism from its near neighbour, Japan, and in turn, much later, was a partial parallel to subsequent Indian tourism. More distinctively, however, Chinese tourism was boosted by VFR and WHM tourism and further differentiated in its pioneering use of smartphone technology. As tourists increasingly came from younger cohorts, marketing and planning for the future became more involved but more difficult. Increasingly cosmopolitan, adventurous and with English-speaking ability, tourists sought greater diversity. An emerging wealthier middle class, millennials with English language capability and technological skills, could largely ignore formal tourist itineraries and package tours, and instead use online blogs and new social media platforms (notably Little Red Book and *Mafengwo*) that enabled eWOM recommendations. In any case, language barriers became less troublesome as smartphones enabled instant translation. Without familiarity with western mores, and without parents as precedents, that was invaluable. Even so, tourism remained anchored to its metropolitan origins where security and shopping were always welcome.

Almost all Chinese tourists now use smartphones, indeed seemingly throughout their waking hours, initially to obtain information about tourist sites, hotels, restaurants and accessibility, as market issues dominated, but smartphones now broke language barriers (evident in shopping practices) giving agency to female tourists and enabling as never before an easy and informed local mobility and instantaneity. Phones created networks of virtual trusted tourist friends, shaped tourist preferences and extended the range of possibilities in destinations, beyond those of conventional marketing strategies and once-popular guidebooks. Photography went from being a memory aid to a means of sharing, and suggesting, experiences.

The smart tourism of social media and photography, adopted by younger tourists, who explored new forms and new places of tourism: more active and more engaged in tasting wine or riding waves, and tentatively venturing beyond the east coast. From the Quad at Sydney University, through wedding photography and dreams of Harry Potter, a novel form of tourism eventually resulted in the creation of several distinctive 'marked places', largely unknown to a wider world. Such colourful sites, and colourful tourists, in and far beyond Sydney, represented a new aesthetics of tourism spearheaded by young middle-class women, empowered through their posts on *xiaohongshu*. Such places, where nature, water and open space almost always played a part, were sites of photography, performativity and fashion, that incidentally generated and shaped a new tourist geography, as they were recommended and identified as *daka* sites. In the late 2010s, *daka* tourism and colourful Hanfu travels with their implications for photography have been adopted by female millennials: thoroughly and uniquely Chinese phenomena.

The significance of these new sites, albeit for an almost exclusively Chinese audience, demonstrates the power of tourist agency through photography to depict and share new destinations, incidentally demonstrating how the ability of conventional media and marketing agencies to directly shape the gaze and market destinations has declined. Beyond that there was the paradox of amateur influencers and celebrities in a nominally socialist state. While that emergent geography was unknown to non-Chinese speakers, its impact at Sydney University was enough for it to filter back into 'normative' travel advice.

Yet the creation of a new Chinese tourist geography of Australia, seemingly a far cry from the package tours of standard iconic sites and animals of less than a decade earlier, was grafted on to the familiar tourism geography of the city. As photographing the Opera House from the vantage point of Lavender Bay indicated, icons were irreplaceable. Despite the 'new' tourism geography offering a spontaneity beyond more tired and predictable package tours, here too newer visitors followed earlier ones, while following Chinese precedents (Ren, 2022). The packages themselves were becoming smaller, and more flexible, while still accounting for the majority of visitors (especially on the east coast with its dominance of direct airline fights) who, with limited time available, sought convenience, value and security, so enabling Chinese businesses to thrive and Chinatowns to prosper. Indeed, for all the new diversity, technological innovation and the empowerment of female tourists, a familiar core remained.

Despite inertia at the core, in no more than a decade Chinese tourism in Australia had made a trajectory from quite conventional packaged mass tourism to something unique and technologically sophisticated: a trajectory unlike any other. Yet there remain gaps and silences. Much of what we have discussed here has been centred around, and devolved from, a formal study of less than a hundred tourists in an unusual site in central Sydney. There is a dearth of information on what Chinese tourists may or may not be involved in, within and beyond the metropolitan fringes, though colourful sites like the Pink Lake have exerted a pull. Beyond rare case studies, few accounts have examined the particular meaning of tourism for the tourists, whether as a more philosophical practice or as a desire for value for money. Whether the 'same' environments are experienced differently, and why some 'marked places' are attached particular significance, is largely unknown. This book has been merely one starting point.

Post-COVID-19. Plagues and Politics

Early in 2020, COVID-19 brought an abrupt end to international tourism, with Chinese tourism in Australia an early victim, ironically at a time when tourist numbers were greatest and future projections extremely positive. It also coincided with new geopolitical tensions between Australia and China centred on questions surrounding the genesis of the virus and spilling over into trade sanctions. COVID-19 stalled the global tourism economy. The decline in tourism between 2020 and 2021 is said to 'have cost the global economy more than $4 trillion' (Fu & Sutunyarak, 2023, p. 163). For three years direct flights between China and Australia were effectively non-existent, while in 2023, although airline connections were re-established, many airlines had reduced plane, staff and support capacity.

By late 2022 Australia had settled in to a more or less post-COVID position where cases were much fewer, hospitals were less overwhelmed, the new version of the virus somewhat less lethal, mortality rates falling, and the country anxious to escape all the constraints to mobility, return to economic growth and re-establish the tourism economy, so vital in many areas (and barely propped up by domestic tourism). However, China was experiencing a new COVID phase, with lockdowns in several cities, and minimal mobility, that continued into 2023. Although the government then started issuing more passports to Chinese citizens, the virus only started to weaken around April. Belatedly that marked a return to a considerable degree of normality in both countries.

Anticipating a revival, when China reopened its border in January 2023, the big three state-owned Chinese airlines announced flights to Australian cities, initially on limited schedules but soon after providing daily services between the two countries (Ashton, 2023). Other airlines, such as Xiamen, followed. Nonetheless, unsurprisingly, in February 2023 flights between China and Australia were still only one-fifth of pre-pandemic capacity. That too soon changed. Geopolitical obstacles remained. Australia was cautious about awarding visas, and denied some seemingly suitable applicants, and China had removed Australia from the approved ADS list. (In January e-commerce marketer Tianni Ren began planning a team-building trip to Australia for her 14 Hangzhou staff, having hoped to visit the pink lakes she had seen on social media. Instead took her group to New Zealand: 'We asked our tour agent but were told that Australia was not on the group tour list. It is a pity that we did not get to see the pink lakes' [ABC, 18 April 2023]). Politics remained problematic, potential group tours were waiting in the wings, and were being influenced by social media, but interest in Australia remained vibrant.

Throughout the worst of the COVID-19 era enthusiasm for travel remained and interest in visiting Australia continued. With COVID literally and metaphorically up in the air, 'safety, short-distance, and companionship [we]re the focal points for outbound travel' (Hao, 2021) were it to restart. A survey of potential travellers predictably found that 83% of respondents to a China Tourism Academy survey wanted a destination without COVID infections: some 72% were reluctant to visit Australia due to uncertainties about infections and the fear of becoming sick and trapped (Hao, 2021). At a time when the virus was highly contagious in both countries it is remarkable that as many as 28% were willing to travel (though that willingness could not be tested). In the absolute midst of the pandemic with all its uncertainties, potential Chinese visitors were as interested as any, in becoming tourists again. It was anticipated that the role of media would be pivotal to shaping and predicting future trends in tourism preferences, and that the perception of protective measures against COVID-19, and its perceived seriousness in different destinations would be keys to any revivals, so that virtual tourism would be a realistic alternative (Wadhar et al., 2023). Other studies have pointed to continued enthusiasm for travel in the right circumstances for many people who were simply waiting until the time was ripe to revert to what was becoming a familiar practice despite, domestically at least, conservatively favouring nature and places close to home (e.g. Jin et al., 2021; Zhang et al., 2022). Locally, within Australia, entrepreneurs were as anxious as any for a revival, especially in Chinatown, where several significant businesses, including prominent restaurants, had collapsed during the main COVID era. With Australia widely seen as a safe and secure destination, prospects of revival seemed relatively positive, at least in theory.

Yet, by early 2023, based on limited numbers of flights (and high fuel prices and ticket costs) Tourism Australia estimated that a return to the 2019 level of 1.4 million Chinese tourists visiting Australia each year, would not occur until 2026 (Bagshaw, 2023). Alongside the economics of travel, and with ADS still to be restored, Chinese-Australian trade disputes (with Chinese bans on Australian coal, barley and wine), and tensions over the status of Taiwan, all raised uncertainties for China-Australia diplomatic relations, and for a full revival of outbound travel from China to Australia (Lade & Frew, 2022). More than at any time in the recent past tourism was embroiled in geopolitics, as it retained crucial elements of soft power (Tse, 2013) thus requiring 'orderly guidance' by the Chinese state (Dai et al., 2017). Even so, by mid-2023, there was little doubt that a new boom, at least in independent travel, was emerging. Ongoing observations at the Quad and beyond showed that the Harry Potter building and other Sydney *daka* sites were becoming as popular as ever.

The future of Chinese tourism to Australia, or anywhere else, became less uncertain. Of course, a new strain of COVID-19 may return with a vengeance or a new unrelated pandemic take its place. With increasing flight numbers and over 57,000 arrivals from China in January and February compared with just over 89,000 visitors for the whole of 2022 (ABS, 2023), even if few were primarily tourists, optimism was possible. In May numbers had increased again but were still only a third of comparable numbers from pre-COVID years, a stark reminder of the lost group tours. More independent travellers were a bonus. At the same time political tensions between Australia and China have had major consequences for trade and migration, and inevitably for tourism. In a rapidly changing world, understanding recent and contemporary Chinese tourism requires an understanding of delicate but crucial shifts in the geopolitical order, that may or may not bring a new tourist order, in a post-COVID world.

Dwelling on Dreams?

> 'It does not do well to dwell on dreams and forget to live' (Albus Dumbledore).

In *Harry Potter and the Sorcerer's Stone*, Professor Albus Dumbledore, the headmaster of Hogwarts School of Witchcraft and Wizardry, finds Harry Potter gazing into the Mirror of Erised, a mirror that shows viewers their deepest desires. He advises Harry not merely to dream, an invocation that President Xi would have supported. That advice may have relevance for the future of Chinese tourism in Australia. The wishes of many people in China and Australia are for a return not only to the actual experience of the era prior to COVID-19, but to what was then projected as a future of rapid and seemingly endless growth in Chinese tourism. Is this likely? The revival of tourism in 2023 emphasised that independent Chinese travellers continued to favour Australia, and that VFR was strengthening this trend. At the University of Sydney, Harry Potter was continuing to attract youthful tourists, and the emergent Chinese geography of tourism was being re-established. The University of Sydney had even lent its Great Hall to local Hogwarts events. Even were newer generations in China to lose their relationship with Harry Potter, it seems that the University will perpetuate the myth for a local market. Perhaps, like Elvis Presley in Parkes, where a long-dead performer who had never been to Australia, let alone to the small NSW country town of Parkes, remains the continued focus of an annual Elvis Presley Festival (Connell & Gibson, 2017), the supposed legacy of Harry Potter at the University of Sydney may later attract others, even those with no interest in Harry Potter at all, but simply because it is there. Campus tourism will continue because the education tourism market, scouting prestigious institutions for a highly successful offspring, is independent of the Harry Potter phenomenon, despite their convergence at the University of Sydney.

Whether or not Harry Potter retains his significance (or perhaps whether future tourists may visit the site where Harry Potter fans once photographed and performed), Australia will surely retain its attraction to Chinese tourists. Koalas, kangaroos and the Opera House will remain drawcards. VFR tourism may well increase. Group tours will take different forms. New niches will follow and marketing, alongside improved transport and accommodation, may take more people beyond the east coast. Colourful niches abound. Geopolitical uncertainties will fluctuate, economic trends will vary and inflation and recession challenge households. China may favour destinies with Belt and Road Initiative countries, and abandon AUKUS coalition countries, or simply discourage outbound tourism just as it has discouraged stand-up comedy and limited the availability of foreign films. But tourism is always subject to politics, economics and increasingly technological change. It lives and perseveres with uncertainty and the whims of fashion.

While the new growth in 2023 may eventually produce a structure of tourism little different from earlier times, similar in many ways to the pre-pandemic tourism we explore in this book, that seems unlikely. It might revive in new forms, with global debates over establishing new clean, green and sustainable tourism, and discouraging pollution from plane emissions. It will change in the post-pandemic era due to geopolitical factors, demographic and educational considerations, and even to trends in the fashion industry. Men may take charge of smartphones, and post quite different blogs and pictures from different sites. Some messaging platforms may disappear (or be disappeared). Smart tourism and further developments in visual telephony, AI and virtual reality travel (where security is guaranteed and costs are low) may obviate the need to travel at all. That too seems presently unlikely, but so did smartphones two decades ago. It does, however, seem likely that the tourism industry, despite its vulnerability to external shocks, will retain its adaptiveness and resilience to external pressures and the shifting agency of tourists. It has had to do so.

These changes in an economic and culturally important industry will necessitate ongoing research in order to understand the evolution of Chinese tourism. More qualitative research, ideally co-produced with Chinese tourists, will generate new knowledge and minimise cultural misunderstandings. While there will always be a need for large-scale quantitative demographic and economic data, it is the diverse lived experiences of people that underpin tourism and cooperation. Sometimes it seems all quite simple; as one tourist said, he merely wanted 'to sing and dance with the local people'. Understanding the various factors that influence Chinese tourism, and awareness of the possibility of new factors emerging in the coming years, is necessary to ensure that research is contextual, nuanced and relevant. We hope that this book is one small contribution and an inspiration to others to travel further, develop themes in more detail and even contribute to mutual understanding.

Bibliography

ABC. (2023, April 18). Chinese tourism to Australia drops dramatically compared to pre-COVID numbers, despite end of travel bans. *ABC News.*

Ashton, C. (2023, January 20). China's 'big three' airlines boost flights to Australia. *Executive Traveller.* https://www.executivetraveller.com/news/china-big-three-airlines-increase-flights-sydney-melbourne

Australian Bureau of Statistics. (2023). *Overseas arrivals and departures, Australia: Statistics on international travel arriving in and departing from Australia.* Australian Bureau of Statistics (ABS).

Bagshaw, E. (2023, February 28). Tourism Australia launches campaign targeting Chinese tourists. *Sydney Morning Herald.*

Connell, J., & Gibson, C. (2017). *Outback Elvis. The story of a festival, its fans and a town called Parkes.* New South.

Dai, B., Jiang, Y., Yang, L., & Ma, Y. (2017). China's outbound tourism – Stages, policies and choices. *Tourism Management, 58*, 253–258.

Fu, H., & Sutunyarak, C. (2023). Demographic factors affecting Chinese tourists traveling to Thailand in the post-Covid-19 era. *Innovative Marketing, 19*(1), 162–174.

Hao, H. (2021, November 25). *Annual report on China's outbound tourism development 2021 released.* Visit Singapore. Org. https://www.visitsingapore.org/global-travel-news/annual-report-on-chinas-outbound-tourism-development-2021-released/

Jin, X., Bao, J., & Tang, C. (2021). Profiling and evaluating Chinese consumers regarding post-COVID-19 travel. *Current Issues in Tourism, 25*(5), 745–763.

Lade, C., & Frew, E. (2022). The future of Chinese outbound tourism. An Australian perspective. In C. Lade & E. Frew (Eds.), *Chinese outbound tourist behaviour* (pp. 180–191). Routledge.

Ren, L. (2022). Chinese tourists' changing behavior in package tours: The suppliers' account. *Tourist Studies, 22*(4), 328–347.

Tse, T. (2013). Chinese outbound tourism as a form of diplomacy. *Tourism Planning and Development, 10*(2), 149–158.

Wadhar, S., Shahani, R., Zhou, R., Siddiquei, A., et al. (2023). What factors will influence Chinese international traveling for leisure in the post-COVID-19 era: Role of health priorities and health-related information literacy. *Healthcare, 11*(3), 315.

Zhang, H., Weng, S., & Bao, J. (2022). How do Chinese tourists negotiate the constraints of engaging in post-COVID-19 domestic travel? *Annals of Tourism Research Empirical Insights, 3*(2), 100065.

Bibliography

ABC. (2016, February 8). Northern Territory tourism campaign targets rich Chinese, offers bragging rights, Sydney. *ABC News*.
ABC. (2019, September 12). Selfies, wedding dresses and campers: China's big crush on Port Gregory's pink lake. *ABC News*.
ABC. (2023, April 18). Chinese tourism to Australia drops dramatically compared to pre-COVID numbers, despite end of travel bans. *ABC News*.
Abidin, C. (2016). Visibility labour: Engaging with influencers' fashion brands and #OOTD advertorial campaigns on Instagram. *Media International Australia, 161*(1), 86–100.
Abidin, C. (2018). *Internet celebrity: Understanding fame online*. Emerald.
Albino, S. (2015). *Tourism in University Cities: The role of universities in place branding*. PhD thesis, University of Exeter and Universidade de Lisboa. https://ore.exeter.ac.uk/repository/handle/10871/21034
Anderson, K. (1990). 'Chinatown Re-oriented': A critical analysis of recent redevelopment schemes in a Melbourne and Sydney enclave. *Australian Geographical Studies, 18*(2), 137–154.
Anderson, K., Ang, I., Del Bono, A., McNeill, D., & Wong, A. (2019). *Chinatown unbound: Trans-Asian urbanism in the age of China*. Routledge.
Ang, I., & Del Bono, A. (2018, May 11). Sydney's Chinatown is much more of a modern bridge to Asia than a historic enclave. *The Conversation*.
Anon. (2019, August 17). Daka destinations. *The Economist*, 30.
Anon. (2022). China. In *The statesman's yearbook 2023* (pp. 303–324). Palgrave Macmillan.

J. Connell et al., *Chinese Tourism in Australia*,
https://doi.org/10.1007/978-981-97-2477-2

Ashton, C. (2023, January 20). China's 'big three' airlines boost flights to Australia. *Executive Traveller*. https://www.executivetraveller.com/news/china-big-three-airlines-increase-flights-sydney-melbourne

Aubusson, L. (2015, August 10). Bare Island drowning sparks warning to public about taking cliff and rock photos. *The Daily Telegraph*.

Austrade. (2017). *Visiting friends and relatives in Victoria: The China and Indian markets*. Austrade.

Austrade. (2019). *Chinese free and independent travellers – Their potential for regional Australia*. Austrade.

Australia Bureau of Statistics. (2018). *3401.0 overseas arrivals and departures, Australia*. ABS.

Australian Bureau of Statistics. (2021). *Migration, Australia. Statistics on Australia's international migration, internal migration (interstate and intra-state), and the population by country of birth*. Australian Bureau of Statistics (ABS).

Australian Bureau of Statistics. (2023). *Overseas arrivals and departures, Australia: Statistics on international travel arriving in and departing from Australia*. Australian Bureau of Statistics (ABS).

Australian Capital Territory Government. (2014). *2020 tourism strategy: Growing the visitor economy 2014–2020*. Australian Capital Territory Government.

Australian Government, Department of Home Affairs. (2022). *Working holiday maker visa program report*. Australian Government, Department of Home Affairs.

Australian Trade and Investment Commission. (2019). *Australia's tourist arrivals exceed 9 million in 2018*. https://www.austrade.gov.au/news/economic-analysis/australias-tourist-arrivals-exceed-9-million-in-2018

Bagshaw, E. (2018, August 5). Ski fields fake it till they make it. *Sydney Morning Herald*, 10.

Bagshaw, E. (2023, February 28). Tourism Australia launches campaign targeting Chinese tourists. *Sydney Morning Herald*.

Banyai, M. (2010). Dracula's image in tourism: Western bloggers versus tour guides. *European Journal of Tourism Research, 3*(1), 5–22.

Bao, J., Jin, X., & Weaver, D. (2019). Profiling the elite middle-age Chinese outbound travellers: A 3rd wave? *Current Issues in Tourism, 22*(5), 561–574.

Becken, S., & Scott, N. (2016, July 8) How Australia can capitalise on Chinese tourism. *The Conversation*.

Beeton, S. (2016). *Film-induced tourism* (2nd ed.). Channel View.

Boag, C., Martin, R., & Bell, D. (2020). An 'authentic' Aboriginal product: State marketing and the self-fashioning of Indigenous experiences to Chinese tourists. *Australian Journal of Anthropology, 31*, 51–65.

Breakey, N., Ding, P., & Lee, T. (2008). Impact of Chinese outbound tourism to Australia. *Current Issues in Tourism, 11*(6), 587–603.

Bronner, F., & de Hoog, R. (2011). Vacationers and eWOM: Who posts and why, where, and what? *Journal of Travel Research, 50*(1), 15–26.

Buckley, R. (2014, April 25). Ecotourism for the Chinese: Up the creek without a paddle. *The Conversation.*

Cai, P. (2017, January 31). Australia must capitalize on Chinese diaspora. *East Asia Forum.*

Campos, A., Mendes, J., Valle, P., & Scott, N. (2018). Co-creation of tourist experiences: A literature review. *Current Issues in Tourism, 21*(4), 369–400.

Carson, S., & Pennings, M. (2020). Online and on tour. The smartphone effect in transmedia contexts. In M. Mansson, A. Buchmann, C. Cassinger, & L. Eskilsson (Eds.), *The Routledge companion to media and tourism* (pp. 414–423). Routledge.

Cash, J., & Kongkunakornkul, P. (2023, April 19). Chinese tourists are back, but numbers still far from pre-COVID levels. *Reuters.* Accessed May 3, 2023, from https://www.reuters.com/world/china/chinese-tourists-are-back-numbers-still-far-pre-covid-levels-2023-04-19/

Chen, H., Weiler, B., & Young, M. (2018). Examining service shortfalls: A case study of Chinese group package tours to Australia. *Journal of Vacation Marketing, 24*(4), 371–386.

Chen, J., Becken, S., & Stantic, B. (2020a). *Identifying Chinese tourists' travel patterns in Australia from Weibo posts.* Unpublished paper to CAUTHE conference, Auckland.

Chen, X., Cheng, Z., & Kim, G. (2020b, 1904). Make it memorable: Tourism experience, fun, recommendation and revisit intentions of Chinese outbound tourists. *Sustainability, 12*(5).

Chen, Z., Suntikul, W., & King, B. (2020c). Research on tourism experiencescapes: The journey from art to science. *Current Issues in Tourism, 23*(11), 1407–1425.

Chen, X., Mak, B., & Calder, I. (2021). The Guimi holidays of young Chinese women: Evidence from Cyber Communities. *Journal of China Tourism Research, 17*(2), 230–248.

Chenyu, L. (2018, July 6). Top Chinese universities cap visitors as campus tourism soars: Peking and Tsinghua universities will only allow a few thousand daily guests as a crowd control measure. *Sixth Tone.* Accessed October 28, 2019, from https://www.sixthtone.com/news/1002576/top-chinese-universities-cap-visitors-as-campus-tourism-soars#

China Tourism Academy. (2017a). *Annual report of China outbound tourism development 2017.* Tourism Education Press. [In Chinese].

China Tourism Academy. (2017b). *Tribute to the Chinese tourists – 2016 China outbound tourists large data* [in Chinese]. Accessed May 3, 2018, from http://www.ctaweb.org/html/2017-1/2017-1-22-10-41-66902.html

China Tourism Academy. (2020). *2019 annual report on China's outbound tourism development, Beijing.* [in Chinese]. Accessed May 3, 2023, from http://eng.ctaweb.org.cn/ctaen/c09/202103/9189e906cd044e0ebcd388693e7fb6ff.shtml

Chiu, C., Ip, C., & Silverman, A. (2012). Understanding social media in China. *McKinsey Quarterly, 2*, 78–81.

Chu, S., Deng, T., & Cheng, H. (2020). The role of social media advertising in hospitality, tourism and travel: A literature review and research agenda. *International Journal of Contemporary Hospitality Management, 32*(11), 3419–3438.

Chung, J., & Buhalis, D. (2008). Web 2.0: A study of online travel community. In P. O'Connor, W. Höpken, & U. Gretzel (Eds.), *Information and communication technologies in tourism* (pp. 70–81). Springer-Wien.

Collins, J., & Jordan, K. (2009). Ethnic precincts as ethnic tourism destinations in urban Australia. *Tourism Culture and Communication, 9*(1–2), 79–92.

Connell, J. (1996). A study of tourism on university campus sites. *Tourism Management, 17*(7), 541–544.

Connell, J. (2000). The role of tourism in the socially responsible university. *Current Issues in Tourism, 3*(1), 1–19.

Connell, J. (2018). 'Timeless charm'? Tourism and development in Southeast Asia. In A. McGregor, L. Law, & F. Miller (Eds.), *Routledge handbook of Southeast Asian development* (pp. 153–168). Routledge.

Connell, J., & Gibson, C. (2017). *Outback Elvis. The story of a festival, its fans and a town called Parkes.* New South.

Connell, J., & Ip, A. (1981). The Chinese in Sydney: From Chinatown to Suburbia. *Asian Profile, 9*(4), 291–308.

Connell, J., & McManus, P. (2019). Flower viewing from horseback? New directions in Chinese tourism in Australia. *Australian Geographer, 50*(3), 333–347.

Connell, J., & Rugendyke, B. (2010). Creating an authentic tourist site? The Australian standing stones, Glen Innes. *Australian Geographer, 41*(1), 87–100.

Connell, J., Ding, X., McManus, P., & Gibson, C. (2021). Social media, popular culture and 'soft heritage': Chinese tourists in search of Harry Potter. *Tourist Studies, 21*(4), 572–595.

Croce, V. (2016). *Tourism in focus: The Chinese outbound travel market.* European Commission.

Crouch, D., Jackson, R., & Thompson, F. (Eds.). (2005). *The media and the tourist imagination: Converging cultures.* Routledge.

Dai, B., Jiang, Y., Yang, L., & Ma, Y. (2017). China's outbound tourism – Stages, policies and choices. *Tourism Management, 58*, 253–258.

Darragh, M. (2009). A whole new world: Companies are learning how to reach Chinese consumers cost-effectively via social media tools. *Communication World, 26*(6), 35.

Destination NSW. (2018). *Destination NSW. Annual report 2017–2018.* Destination NSW.

Destination NSW. (2019). *Mainland China market profile. Year ended December 2018. For the period of January 2018 to December 2018.* Destination NSW. https://www.destinationnsw.com.au/wp-content/uploads/2019/04/mainland-china-ye-dec-18.pdf?x15361

Destination NSW. (2020). *Sydney precinct visitor profiles.* https://www.destinationnsw.com.au/tourism/facts-and-figures/sydney-tourism-statistics/sydney-precinct-visitor-profiles

Dezan Shira and Associates. (2023). *China briefing: Doing business in China.* https://www.china-briefing.com/doing-business-guide/china/human-resources-and-payroll/china-public-holidays-2021

Dinhopl, A., & Gretzel, U. (2016). Selfie-taking as touristic looking. *Annals of Tourism Research, 57*, 126–139.

Du, X., Liechty, T., Santos, C., & Park, J. (2022). 'I want to record and share my wonderful journey': Chinese Millennials' production and sharing of shortform travel videos on TikTok or Douyin. *Current Issues in Tourism, 25*(21), 3412–3424.

Edensor, T. (1998). *Tourists at the Taj: Performance and meaning at a symbolic site.* Routledge.

Ek Styvén, M., & Foster, T. (2018). Who am I if you can't see me? The 'self' of young travellers as drivers of eWOM in social media. *Journal of Tourism Futures, 4*(1), 80–92.

Erni, J. (2008). Enchanted: Harry Potter and magical capitalism in urban China. *Chinese Journal of Communication, 1*(2), 138–155.

Erni, J. (2013). When Chinese youth meet Harry Potter. Translating consumption and middle-class identification. In A. Fung (Ed.), *Asian popular culture: The global (dis)continuity* (pp. 21–41). Routledge.

Ettlinger, N. (2019). Digital presumption, crowdsourcing, and the digital regime of work. *New Formations, 99*, 52–65.

Eusébio, C., Carneiro, M., Madaleno, M., Robaina, M., et al. (2021). The impact of air quality on tourism: A systematic literature review. *Journal of Tourism Futures, 7*(1), 111–130.

Evlin, L. (2019, March 4). The 'magical' reason Chinese tourists are flocking to the University of Sydney. *SBS News.*

Fang, Y., Chien, P., & Walters, G. (2021). Understanding the emerging Chinese Working Holiday market. *Journal of Vacation Marketing, 27*(1), 3–16.

Feiyue, Y. (2017, March 23). Overture to tourists hits the high notes. *China Daily.*

Fotis, J., Buhalis, D., & Rossides, M. (2011). Social media impact on holiday travel planning: The case of the Russian and the FSU Markets. *International Journal of Online Marketing, 1*(4), 1–19.

Fredline, L., & Jin, X. (2017). The attractiveness of Australian events to Chinese visitors and barriers to attendance. *Event Management, 21*(6), 697–711.

Freed, J. (2016, March 18). Young Chinese prepared to spend big on travel. *Sydney Morning Herald*, 19.

Fu, H., & Sutunyarak, C. (2023). Demographic factors affecting Chinese tourists traveling to Thailand in the post-Covid-19 era. *Innovative Marketing, 19*(1), 162–174.

Gao, J., & Zhang, L. (2017). The changing Chinese community mediascape since the early 1990s. In J. Budarick & G.-S. Han (Eds.), *Minorities and media: Producers, industries, audiences* (pp. 59–84). Macmillan.

Gao, J., Zeng, X., Zhang, C., & Porananond, P. (2022). Understanding the young middle-class Chinese outbound tourism consumption: A social practice perspective. *Tourism Management, 92*, 104555.

Gardiner, S. (2021). Chinese learn to surf: Learning ambidexterity and tourism experience innovation. *Journal of China Tourism Research, 17*(1), 142–162.

Gardiner, S., & Kwek, A. (2017). Chinese participation in adventure tourism: A study of Generation Y International Students' perceptions. *Journal of Travel Research, 56*(4), 496–506.

Garner, A. (2020). I came, I saw, I selfied. Travelling in the age of Instagram. In M. Mansson, A. Buchmann, C. Cassinger, & L. Eskilsson (Eds.), *The Routledge companion to media and tourism* (pp. 316–325). Routledge.

Gibson, C. (2018). Tourism nation: Changing meanings of Cultural Tourism. In G. Turner, D. Rowe, & E. Waterton (Eds.), *Making culture. Commercialisation, transnationalism, and the state of 'nationing' in contemporary Australia* (pp. 116–128). Routledge.

Gibson, C., & Connell, J. (2005). *Music and tourism. On the road again.* Channel View.

GMA. (2022). *How to get started on the Chinese App Little Red Book aka RED as a business?* https://marketingtochina.com/chinese-app-little-red-book/

Gothie, S. (2016). Playing 'Anne': Red braids, Green Gables, and literary tourists on Prince Edward Island. *Tourist Studies, 16*(4), 405–421.

Gray, D. (2018, July 25). China's demand is a corker for wine as records tumble. *Sydney Morning Herald*, 20–21.

Gretzel, U., & Jamal, T. (2009). Conceptualising the creative tourist class: Technology, mobility and tourism experiences. *Tourism Analysis, 14*(4), 471–481.

Grindlay, D. (2016, March 24). Chinese tourists flood isolated grain town, throwing drought-stricken community an unexpected lifeline. *ABC Rural.*

Gu, Q., & Huang, S. (2019). Profiling Chinese wine tourists by wine tourism constraints: A comparison of Chinese Australians and long-haul Chinese tourists in Australia. *International Journal of Tourism Research, 21*(2), 206–220.

Guan, Z. (2021). Chinese beauty bloggers: Amateurs, entrepreneurs, and platform labour. *Celebrity Studies, 12*(2), 326–332.

Guilliatt, R. (2020, January 25). Invasion of the insta tourists. *Weekend Australian Magazine*, 20–23.

Guo, J. (2022). The postfeminist entrepreneurial self and the platformisation of labour: A case study of yesheng female lifestyle bloggers on Xiaohongshu. *Global Media and China, 7*(3), 303–318.

Gupta, N., & Ray, A. (2022). Probing 'instaworthiness': Siting the selfie. *Convergence, 28*(3), 681–698.

Haiyan, C. (2010). An Impact of social media on online travel information search in China. In *3rd international conference on information management, innovation management and industrial engineering* (Vol. 3, pp. 509–512).

Han, M. (2016, February 10). Million is just a beginning, prosperous will continue for years. *Chinese New Express Daily*, 2. [In Chinese].

Han, X., & Cheer, J. (2018). Chinese tourist mobilities and destination resilience: Regional tourism perspectives. *Asian Journal of Tourism Research, 3*(1), 159–187.

Hao, H. (2021, November 25). *Annual report on China's outbound tourism development 2021 released.* Visit Singapore. Org. https://www.visitsingapore.org/global-travel-news/annual-report-on-chinas-outbound-tourism-development-2021-released/

Hao, F., Zhang, S., & Xiao, H. (2021). A return to innocence: Guimi tourism and women empowerment. *International Journal of Tourism Research, 23*(6), 971–983.

Harwit, E. (2017). WeChat: Social and political development of China's dominant messaging app. *Chinese Journal of Communication, 10*(3), 312–327.

Hayward, T. (2020, July 4–5) Cambridge's new tourist problem. *Financial Times*, 14.

Hjorth, L., Burgess, J., & Richardson, I. (2012). *Studying mobile media: Cultural technologies, mobile communication, and the iphone.* Routledge.

Hobsbawm, E., & Ranger, T. (1983). *The invention of tradition.* Cambridge University Press.

Hooper, F. (2017, January 11). Chinese newlyweds make the most of Tasmania's blooming lavender. *ABC Northern Tasmania.*

Horowitz, J. (2019, August 6). The bride, the groom and the Greek sunset: A perfect wedding. *New York Times.*

Hsu, C., & Huang, S. (2016). Reconfiguring Chinese cultural values and their tourism implications. *Tourism Management, 54*, 230–242.

Huang, S., & Chen, G. (2016). Current state of tourism research in China. *Tourism Management Perspectives, 20*, 10–18.

Huang, S., & Gross, M. (2010). Australia's destination image among Mainland Chinese travelers: An exploratory study. *Journal of Travel and Tourism Marketing, 27*, 63–81.

Huang, S., & van der Veen, R. (2019). The moderation of gender and generation in the effects of perceived destination image on tourist attitude and visit intention: A study of potential Chinese visitors to Australia. *Journal of Vacation Marketing, 25*(3), 375–389.

Hughes, G. (1992). Tourism and the geographical imagination. *Leisure Studies, 11*(1), 31–42.

Hwang, Y. (2011). A theory of unplanned travel decisions: Implications for modeling on-the-go travelers. *Information Technology and Tourism, 12*, 283–296.

IPSOS. (2016). *Chinese outbound tourist consumption 2015–2016*. https://www.ipsos.com/sites/default/files/2016-06/021.1-Chinese-Outbound-Tourist.pdf

Ironside, R. (2016, March 18). Chinese tourism warning. *Daily Telegraph*, 21.

Jang, K. (2020). Creating the sacred places of pop culture in the age of mobility: Fan pilgrimages and authenticity through performance. *Journal of Tourism and Cultural Change, 18*(1), 42–57.

Jang, K., & Yamamura, T. (2020). Creative fandoms and the mediatized sacred sites. In M. Månsson, A. Buchmann, C. Cassinger, & L. Eskilsson (Eds.), *The Routledge companion to media and tourism* (pp. 307–315). Routledge.

Jia, Y. (2017). A review of tourist behaviour studies on social media, based on overseas literature. *Tourism Tribune, 32*(4), 117–126. [In Chinese].

Jin, H., & Cai, W. (2022). Understanding the smartphone usage of Chinese outbound tourists in their shopping practices. *Current Issues in Tourism, 25*(18), 2955–2968.

Jin, X., & Sparks, B. (2017). Barriers to offering special interest tour products to the Chinese outbound group market. *Tourism Management, 59*, 2015–2215.

Jin, X., Wu, L., Becken, S., & Ding, P. (2016). How do worry, self-efficacy and coping interact? Examining Chinese tourists to Australia. *Journal of China Tourism Research, 12*(3–4), 374–393.

Jin, H., Moscardo, G., & Murphy, L. (2020). Exploring Chinese outbound tourist shopping: A social practice framework. *Journal of Travel Research, 59*(1), 156–172.

Jin, X., Bao, J., & Tang, C. (2021). Profiling and evaluating Chinese consumers regarding post-COVID-19 travel. *Current Issues in Tourism, 25*(5), 745–763.

Jing, S. (2017, January 13). For the rich, the only class is first class. *China Watch*, 3.

JLL. (2019, January 18). *Australian luxury hotels up their game for Chinese tourists*. Accessed April 22, 2022, from https://www.jll.com.co/en/trends-and-insights/cities/australian-luxury-hotels-game-chinese-tourists

Kaplan, S. (2023, January 27). WeChat tightens Beijing's grip on Chinese diaspora. *Wall Street Journal*.

Kashiwagi, S., Nagai, H., & Furutani, T. (2018). VFR travel generated by international students: The case of Japanese students in Australia. *Tourism: An International Interdisciplinary Journal, 66*(1), 89–103.

Keane, M. (2017). Foreword. In M. Kent, K. Ellis, & J. Xu (Eds.), *Chinese social media. social, cultural, and political implications* (pp. 1–3). Routledge.

Kim, S. (2012). Audience involvement and film tourism experience: Emotional places, emotional experiences. *Tourism Management, 33*(2), 387–396.

Kim, J., & Fesenmaier, D. (2017). Sharing tourism experiences: The post trip experience. *Journal of Travel Research, 56*(1), 28–40.

Kim, S., Guo, Y., & Agrusa, J. (2005). Preference and positioning analyses of overseas destinations by Mainland Chinese outbound pleasure tourists. *Journal of Travel Research, 44*, 212–220.

Kimber, S., Yang, J., & Cohen, S. (2019). Performing love, prosperity and Chinese hipsterism: Young independent travellers in Pai, Thailand. *Tourist Studies, 19*(2), 164–191.

King, R., Racherla, P., & Bush, V. (2014). What we know and what we don't know about online Word-of-Mouth: A review and synthesis of the literature. *Journal of Interactive Marketing, 28*, 167–183.

Kohler, A. (2017, October 28). Daigou shoppers ring up the sales. *The Australian*.

Kommonen, K. (2011). Narratives on Chinese colour culture in business contexts: The Yin Yang Wu Xing of Chinese values. *Cross Cultural Management, 18*(3), 366–383.

Kong, S., Guo, J., & Huang, D. (2022). The girlfriend getaway as an intimacy. *Annals of Tourism Research, 92*, 103337.

Koob, S. (2017, October 28). China girls drive the Aussie tourism boom. *The Australian*.

Kwek, A., & Lee, Y. (2008). Intra-cultural variance of Chinese tourists in destination image project: Case of Queensland, Australia. *Journal of Hospitality and Leisure Marketing, 16*(1–2), 105–135.

Lade, C., & Frew, E. (2022). The future of Chinese outbound tourism. An Australian perspective. In C. Lade & E. Frew (Eds.), *Chinese outbound tourist behaviour* (pp. 180–191). Routledge.

Larsen, J. (2005). Performativity of tourist photography. *Space and Culture, 8*(4), 416–434.

Larsen, K. (2022). (Re)claiming Harry Potter fan pilgrimage sites. In L. Brenner (Ed.), *Playing Harry Potter: Essays and interviews on fandom and performance* (pp. 38–54). McFarland.

Le Heron, E. (2020). Assembling film-landscape relations: Performative practices of The Lord of the Rings tours. *New Zealand Geographer, 76*(1), 39–48.

Leaver, K., & De Poloni, G. (2019, June 30). How a little blue boathouse in Perth became the hottest selfie spot on Instagram. *ABC News*. Accessed August 10, 2023, from https://www.abc.net.au/news/2019-06-30/perth-blue-boathouse-in-crawley-becomes-instagram-star/11261312

Lee, C. (2012). 'Have magic, will travel'. Tourism and Harry Potter's United (Magical) Kingdom. *Tourist Studies, 12*(1), 52–69.

LEK. (2016). *Enter the Dragon: How the China tourism boom is transforming Australia*. Australia China Business Council.

LEK. (2017). *The China tourism economy: Reaching Australia's potential.* Australia China Business Council.

Leung, D., Law, R., van Hoof, H., & Buhalis, D. (2013). Social media in tourism and hospitality: A literature review. *Journal of Travel and Tourism Marketing, 30*(1–2), 3–22.

Leung, D., Li, G., Fong, L., Law, R., & Lo, A. (2014). Current state of China tourism research. *Current Issues in Tourism, 17*(8), 679–704.

Levy, M. (2015, August 10). Drowning victim was posing for photo when swept off rocks at La Perouse. *The Sydney Morning Herald.* Accessed February 25, 2020, from https://www.smh.com.au/national/nsw/woman-drowns-after-being-swept-off-rocks-at-la-perouse-20150809-giv2jv.html

Lexhagen, M., Larson, M., & Lundberg, C. (2013). The virtual fan(g) community: Social media and pop culture tourism. In *Tourism social media: Transformations in identity, community and culture* (pp. 133–157). Emerald.

Li, W. (2013) *Research on the development of campus tourism – Taking the universities of China and Britain as an example.* PhD thesis, Nanchang University.

Li, X. (2016, February 5). During the Spring Festival, a large number of Chinese tourists are expected to visit Australia. *Australian New Express Daily*, 1.

Li, F. (2020). Understanding Chinese tourists' motivations of sharing travel photos in WeChat. *Tourism Management Perspectives, 33*, 100584.

Li, J., & Carr, N. (2004). Visitor satisfaction: An analysis of mainland Chinese tourists on the Australian Gold Coast. *International Journal of Hospitality and Tourism Administration, 5*(3), 31–48.

Li, J., Pearce, P. L., & Low, D. (2018). Media representation of digital-free tourism: A critical discourse analysis. *Tourism Management, 69*, 317–329.

Li, M., Sharpley, R., & Gammon, S. (2019). Towards an understanding of Chinese tourist photography: Evidence from the UK. *Current Issues in Tourism, 22*(5), 505–521.

Lin, X. (2022). Domestic tourism on the rise. *China Daily.* https://global.chinadaily.com.cn/a/202201/01/WS61cfbe40a310cdd39bc7ea93_1.html

Lin, Z., & Yang, L. (2021). Smartphones as actors: A new digital disability care actor-network in China. *International Journal of Cultural Studies, 24*(4), 673–688.

Ling, L., & Dai, Y. (2012). Discussion on the sustainable development of campus tourism. *Market Forum, 6*, 94–95.

Litvin, S., Goldsmith, R., & Pan, B. (2008). Electronic word-of-mouth in Hospitality and Tourism Management. *Tourism Management, 29*(3), 458–468.

Liu, C. (2021). Exploring selfie practices and their geographies in the digital society. *Geographical Journal, 187*, 240–252.

Liu, C. (2022). Imag(in)ing place: Reframing photography practices and affective social media platforms. *Geoforum, 29*, 172–180.

Liu, B., & Pennington-Gray, L. (2015). Bed bugs bite the hospitality industry? A framing analysis of bed bug news coverage. *Tourism Management, 48*, 33–42.

Liu, Y., Bao, J., & Chen, K. (2017). Sentimental features of Chinese outbound tourists in Australia: Big-data based content analysis. *Tourism Tribune, 32*, 46–58. [In Chinese].

Liu, Y., Huang, K., Bao, J., & Chen, K. (2019). Listen to the voices from home: An analysis of Chinese tourists' sentiments regarding Australian destinations. *Tourism Management, 71*, 337–347.

Liu, X., Mehraliyev, F., Liu, C., & Schuckert, M. (2020). The roles of social media in tourists' choices of travel components. *Tourist Studies, 20*(1), 27–48.

Liu, X., Wang, D., & Gretzel, U. (2022). On-site decision-making in smartphone-mediated contexts. *Tourism Management, 88*, 104424.

Lonely Planet. (2017). *New South Wales*. Lonely Planet.

Long, F., Aziz, N., & Ngah, A. (2023). Exploring the relationship between travel motivations and information sharing behaviors on social media: Gendered differences of Chinese Gen Y in a context of Confucianism. *Journal of China Travel Research, 19*(2), 245–265.

Loy-Wilson, S. (2017). Coolie alibis: Seizing gold from Chinese miners in New South Wales. *International Labor and Working Class History, 91*, 28–45.

Lyu, S. (2016). Travel selfies on social media as objectified self-presentation. *Tourism Management, 54*, 185–195.

Ma, Y., & Cai, W. (2023). Social media in ethnographic research: Critical reflections on using WeChat in researching Chinese outbound tourists. *Current Issues in Tourism, 26*(20), 3275–3287.

Ma, E., Liu, Y., Li, J., & Chen, S. (2016). Anticipating Chinese tourists arrivals in Australia: A time series analysis. *Tourism Management Perspectives, 17*, 50–58.

Ma, Y., Hardy, A., & Ooi, C. (2020). Researching Chinese Tourists on the Move. *Journal of China Tourism Research, 16*(2), 214–229.

Madionou, M. (2016). Ambient co-presence: Transnational family practices in polymedia environments. *Global Networks, 16*(2), 183–201.

Magno, F., & Cassa, F. (2018). The impact of social media influencers in tourism. *Anatolia, 29*(2), 288–290.

Mangi, M., Yue, Z., & Kalwar, S. (2019). Universities emerging as tourism sites in China: A case study of Tsinghua University Beijing. *Cogent Social Sciences, 5*, 1–26.

Mannheimer, E., Reijnders, S., & Brandellero, A. (2022). Game of Thrones tourism and the (re)imagination of the new Northern Ireland. *International Journal of Cultural Studies, 25*(2), 554–569.

Marwick, A., & boyd, d. (2011). I tweet honestly, I tweet passionately: Twitter users, context collapse and the imagined audience. *New Media and Society, 13*(1), 114–133.

McCartney, G., & Pek, R. (2018). An examination of Sina Weibo travel blogs' influence on sentiment towards hotel accommodation in Macao. *Journal of China Tourism Research, 14*(2), 146–157.

McHugh, B. (2017, June 15). Australia must be China-ready to capitalise on its unique tourism. *ABC Rural.*

McManus, P., Connell, J., & Ding, X. (2021). Chinese tourists at the University of Sydney: Constraints to co-creating Campus Tourism? *Current Issues in Tourism, 24*(24), 3508–3518.

McWha, M., & Beeton, S. (2018). Engaging with popular media through tourism: A brief introduction. *Tourist Studies, 18*(3), 257–260.

Miao, Y. (2015). The influence of Electronic-Wom on tourists' behavioral intention to choose a destination: A case of Chinese tourists visiting Thailand. *AU-GSB e-JOURNAL, 8*(1), 13–31.

Milazzo, L., & Santos, C. (2022). Fanship and imagination: The transformation of everyday spaces into *Lieux D'Imagination. Annals of Tourism Research, 94*, 103399.

Ming, L. (2007). *Research in Campus Tourism: Actuality, problems and advice – A Literature Review of 10 years research in Campus Tourism in China*. DOI: CNKI:SUN:ZTKB.0.2007–06-030. Accessed August 24, 2019.

Munar, A., & Jacobsen, J. (2014). Motivations for sharing tourism experiences through social media. *Tourism Management, 43*, 46–54.

Murphy, D. (2017, May 13). Middle-income Chinese are the new wave of tourism. *Sydney Morning Herald.*

Ni, J. (2020). *Daka travel experience: An examination of Chinese Millennial travellers.* MSc thesis, Purdue University.

Northern Territory. (2017). *China ready industry program*, mimeo.

NSW Business Chamber. (2017). *NSW visitor economy. On the move mid-term report card.* Sydney.

Nusair, K., Bilgihan, A., Okumus, F., & Cobanoglu, C. (2013). Generation Y travelers' commitment to online social network websites. *Tourism Management, 35*, 13–22.

O'Regan, M., & Choe, J. (2022). Searching for prestige: Motivations and managerial implications of Chinese campus tourists. *Leisure Studies, 41*(6), 862–878.

Oliveira, T., Araujo, A., & Tam, C. (2020). Why do people share their travel experiences on social media? *Tourism Management, 78*, 1–14.

Ooi, C., et al. (2021). *Realising the opportunities of Chinese tourism: A comparative study.* Asia Institute, University of Tasmania.

Osmond, A., Chen, T., & Pearce, P. (2015). Examining experience economy approaches to tourists' anticipated experiences: Mainland Chinese travellers consider Australia. *European Journal of Tourism Research, 10*, 95–108.

Outbound Tourist (City) Consumption, 2015–2016. Beijing: World Tourism Cities Federation.

Pan, G., & Laws, E. (2001). Tourism marketing opportunities for Australia in China. *Journal of Vacation Marketing, 8*(1), 39–48.

Pan, S., Tsai, H., & Lee, J. (2011). Framing New Zealand: Understanding tourism TV commercials. *Tourism Management, 32*(3), 596–603.

Pearce, P., Wu, M., & Osmond, A. (2013). Puzzles in understanding Chinese tourist behaviour: Towards a Triple-C Gaze. *Tourism Recreation Research, 8*(2), 145–157.

Pearce, P., Wu, M., & Chen, T. (2015). The spectacular and the mundane: Chinese tourists' online representations of an iconic landscape journey. *Journal of Destination Marketing and Management, 4*, 24–35.

Peng, Y. (2019). Sharing food photographs on social media: Performative Xiaozi lifestyle in young, middle-class urbanites' WeChat 'Moments'. *Social Identities, 25*(2), 269–287.

Perkins, H., & Thorns, D. (2001). Gazing or performing? Reflections on Urry's tourist gaze in the context of contemporary experience in the antipodes. *International Sociology, 16*(2), 185–204.

Pham, T., Nghiem, S., & Dwyer, L. (2017). The determinants of Chinese visitors to Australia: A dynamic demand analysis. *Tourism Management, 63*, 268–276.

Pham, T., Nghiem, S., & Dwyer, L. (2018). The economic impacts of a changing visa fee for Chinese tourists to Australia. *Tourism Economics, 24*(1), 109–126.

Pourfakhimi, S., Duncan, T., & Coetzee, W. (2020). Electronic word of mouth in tourism and hospitality consumer behaviour: State of the art. *Tourism Review, 75*(4), 637–661.

Prayag, G., Hall, C., & Wood, H. (2018). I feel good! Perceptions and emotional responses of Bed and Breakfast providers in New Zealand toward Trip Advisor. *Journal of Hospitality Marketing and Management, 27*(1), 1–20.

Qi, S., & Chen, N. (2019). Understanding Macao's destination image through User-generated content. *Journal of China Tourism Research, 15*(4), 503–519.

Qu, H., & Lee, H. (2011). Travellers' social identification and membership behaviours in online travel community. *Tourism Management, 32*(6), 1262–1270.

Randwick City Council. (2016, February 11). *Shock measures taken to increase safety for Sydney's rock fishers*. Randwick City Council. Accessed February 25, 2020, from http://www.randwick.nsw.gov.au/about-council/news/news-items/2016/february/shock-measures-taken-to-increase-safety-for-sydneys-rock-fishers

Ren, L. (2022). Chinese tourists' changing behavior in package tours: The suppliers' account. *Tourist Studies, 22*(4), 328–347.

Rough Guides. (2017). *The Rough Guide to Australia* (12th ed.). Rough Guide.

Rowen, I. (2016). The geopolitics of tourism. *Annals of the Association of American Geographers, 106*(2), 385–393.

Rowen, I. (2021). Tourism studies is a geopolitical instrument: Conferences, Confucius Institutes, and 'the Chinese Dream'. *Tourism Geographies, 23*(4), 895–914.

Ruhanen, L., Whitford, M., & McLennan, C. (2015). Exploring Chinese visitor demand for Australia's Indigenous tourism experiences. *Journal of Hospitality and Tourism Management, 24*, 25–34.

Sandel, T., & Wang, Y. (2022). Selling intimacy online: The multi-modal discursive techniques of China's *wanghong*. *Discourse, Context and Media, 47*, 100606.

Sann, R., Lai, P., & Chen, C. (2021). Review papers on eWOM: Prospects for hospitality industry. *Anatolia, 32*(2), 177–206.

Segota, T. (2018). (G)A(i)ming at the throne: Social media and the use of visitor-generated content in destination marketing. In C. Lundberg & V. Ziakas (Eds.), *The Routledge handbook of popular culture and tourism* (pp. 427–438). Routledge.

Senft, T. (2013). Microcelebrity and the branded self. In J. Hartley, A. Bruns, & J. Burgess (Eds.), *New media dynamics* (pp. 346–354). Wiley.

Shi, S. (2021). *Travelling with Hanfu: A social media analysis of contemporary Chinese travelling for artistic photographs*. MIHM thesis, Auckland University of Technology.

Sier, J. (2017, October 6). Daigou retailer sells well on ASX debut. *Sydney Morning Herald*.

Sigala, M. (2016). Social media and the co-creation of tourism experiences. In M. Sotiriadis & D. Gursoy (Eds.), *The handbook of managing and marketing tourism experiences* (pp. 85–111). Routledge.

Skavronskaya, L., Moyle, B., & Scott, N. (2019). Collecting memorable tourism experiences: How do 'wechat? *Journal of China Tourism Research, 16*(3), 424–446.

Smith, M. (2016). Delivering the real experience. *Australian Financial Review*, 42–43.

Song, S. (2022, June 29). RED (Xiaohongshu) trends 2022: emerging trends in fitness, fashion & beauty. *China Digital Times*. https://walkthechat.com/red-xiaohongshu-consumer-trends-2022

Sotiriadis, M., & van Zyl, C. (2013). Electronic word-of-mouth and online reviews in tourism services: The use of twitter by tourists. *Electronic Commerce Research, 13*, 103–124.

South Australian Tourism Commission (2013) Snapshot. (2013). *Activating China – 2020*. Government of South Australia.

Sparks, B., & Pan, G. (2009). Chinese outbound tourists: Underrating their attitudes, constraints and use of information services. *Tourism Management, 30*(4), 483–494.

Spracklen, K., & Spracklen, B. (2014). The strange and spooky battle over bats and black dresses: The commodification of Whitby Goth Weekend and the loss of a subculture. *Tourist Studies, 14*(1), 86–02.
State Government Victoria. (2012). *Victoria's China tourism strategy*. State Government Victoria.
Stylianou-Lambert, T. (2012). Tourists with Cameras: Reproducing or Producing? *Annals of Tourism Research, 39*(4), 1817–1838.
Su, M., Wall, G., Wu, B., et al. (2021). Tourism place making through the bioluminescent 'Blue Tears' of Pingtan Islands, China. *Marine Policy, 133*, 104744.
Sun, W. (2016). *Chinese language media in Australia: Developments, challenges and opportunities*. UTS Australia-China Relations Institute.
Sun, W., Yue, A., Sinclair, J., & Gao, J. (2011). Diasporic Chinese media in Australia: A post-2008 overview. *Continuum, 25*(4), 515–527.
Sun, M., Ryan, C., & Pan, S. (2015). Using Chinese travel blogs to examine perceived destination image: The case of New Zealand. *Journal of Travel Research, 54*(4), 543–555.
Symington, A. (2018). *Sydney*. Lonely Planet.
Tan, S. (2017, October 6). Golden week for tourism, shops. *Australian Financial Review*.
Tan, Y. (2024). 'Capturing the shining moments': Performing liminoid Seichi Junrei on social media. *Current Issues in Tourism, 27*(12), 1959–1972.
Tan, W., & Yang, C. (2021). The relationship between narcissism and landmark check-in behaviour on social media. *Current Issues in Tourism, 24*(24), 3489–3507.
Taylor, D. (2020). Putting the 'self' in selfies: How narcissism, envy and self-promotion motivate sharing of travel photos through social media. *Journal of Travel and Tourism Marketing, 37*(1), 64–77.
Tham, A., Zhang, X., Spinks, W., & Wang, L. (2022). Chinese outbound tourists and *Daigou* as soft power. *Journal of China Tourism Research, 18*(6), 1143–1163.
Thelen, T., & Kim, S. (2023). Understanding fan tourists at a non-commodified fan pilgrimage site: An application of fan capital theory. *Current Issues in Tourism, 26*(1), 12–17.
Thurnell-Read, T. (2017). What's on your bucket list?'. Tourism, identity and imperative experiential discourse. *Annals of Tourism Research, 67*, 58–66.
Tourism and Events Queensland. (2013). *Meeting the expectations of your Chinese visitors and making them feel welcome*. Tourism and Events Queensland.
Tourism and Events Queensland. (2016). *Queensland Asia tourism strategy 2016–2025*. Tourism and Events Queensland.
Tourism Australia. (2011). *China tourism strategy, 2012–2020*. Tourism Australia.
Tourism Australia. (2016). *China market profile 2015*. Tourism Australia.
Tourism Australia. (2017). *China market profile 2016*. Tourism Australia.

Tourism Australia. (2018). *International tourism snapshot as at 31 December 2017*. Tourism Australia.

Tourism Research Australia. (2018). *Trips, nights, average stay and spend by international visitors, international tourism statistics*. Australian Trade and Investment Commission. Accessed May 18, 2018.

Tourism Tasmania. (2016). *Chinese visitor snapshot*. Tourism Tasmania.

Tourism Western Australia. (2011). *Our direction in China 2012–2015*. Tourism Western Australia.

Tse, T. (2013). Chinese outbound tourism as a form of diplomacy. *Tourism Planning and Development, 10*(2), 149–158.

Tse, T., & Tse, Q. (2015). The legal aspects of 'Zero-fare' tour in shopping tourism: A case of Chinese visitors in Hong Kong. *Journal of China Tourism Research, 11*(3), 297–314.

Tung, V., Lee, S., & Hudson, S. (2017). The potential of anime for destination marketing: Fantasies, otaku, and the kidult segment. *Current Issues in Tourism, 22*(12), 1423–1436.

Tussyadiah, I., & Fesenmaier, D. (2009). Mediating tourist experiences: Access to places via shared videos. *Annals of Tourism Research, 36*(1), 24–40.

Urbain, J.-D. (1989). The tourist adventure and his images. *Annals of Tourism Research, 16*(1), 106–118.

Urry, J. (1990). *The Tourist Gaze*. Sage.

Van Hoa, T., Turner, L., & Vu, J. (2018). Economic impact of Chinese tourism on Australia: A new approach. *Tourism Economics, 24*(6), 677–689.

Victoria The Education State. (2018). *Engaging with Chinese tourists*. Victoria State Government.

Wadhar, S., Shahani, R., Zhou, R., Siddiquei, A., et al. (2023). What factors will influence Chinese international traveling for leisure in the post-COVID-19 era: Role of health priorities and health-related information literacy. *Healthcare, 11*(3), 315.

Walsh, M., Dale, N., & Johns, R. (2019, December 27). #travelgram: Live tourist snaps have turned solo adventures into social occasions. *The Conversation*.

Wang, Z., & Connell, J. (2021). Taiwanese working holiday makers in rural and regional Australia: Temporary transnational identities and employment challenges. *Australian Geographer, 52*(2), 191–207.

Wang, Y., & Davidson, M. (2010a). Pre- and post-trip perceptions: An insight into Chinese package holiday market to Australia. *Journal of Vacation Marketing, 16*(2), 111–123.

Wang, Y., & Davidson, M. (2010b). Chinese leisure tourists: Perceptions and satisfaction with Australia. *Tourism Analysis, 14*(6), 737–747.

Wang, D., Park, S., & Fesenmaier, D. (2016). The role of smartphones in mediating the touristic experience. *Journal of Travel Research, 51*(4), 371–387.

Wang, D., Kirillova, K., & Lehto, X. (2020). Tourism mobilities through time in China: A developmental and holistic lens. *Journal of Travel Research, 59*(6), 1073–1090.

Waysdorf, A., & Reijnders, S. (2018a). The role of imagination in the film tourism experience: The case of Game of Thrones. *Participations. Journal of Audience and Reception Studies, 14*(1), 170–191.

Waysdorf, A., & Reijnders, S. (2018b). Immersion, authenticity and the theme park as social space: Experiencing the wizarding world of Harry Potter. *International Journal of Cultural Studies, 21*(2), 173–188.

Wei, Y., & Wu, T. (2021). Visual representation of a linear tourist destination based on social network photos: A comparative analysis of cross-cultural perspectives. *Journal of Tourism and Cultural Change, 19*(6), 781–804.

Wei, X., Meng, F., & Zhang, P. (2017). Chinese Citizens' outbound destination choice: Objective and subjective factors. *International Journal of Tourism Research, 19*, 38–49.

Wei, Y., Zhang, G., & Weng, W. (2018). The influence of 'Internet Plus' on China's tourism: Evolution and breakthrough. In *Proceedings of 4th international conference on behavioral, economic, and socio-cultural computing (BESC) July 1, 2017, BESC* (pp. 1–5).

Wei, S., Ng, S., Lee, J., & Soutar, G. (2023). Similarity-attraction cluster of outbound Chinese tourists: Who belongs there? *Journal of Hospitality & Tourism Research, 47*(1), 33–56.

Weiler, B., & Chen, H. (2016). Repositioning Australian cities as settings for memorable nature-based experiences aimed at Chinese tourists. *Tourism Recreation Research, 41*(3), 246–258.

Weiler, B., & Yu, X. (2006). *Understanding experiences of Chinese visitors to Victoria*. Sustainable Tourism Cooperative Research Centre.

Weiler, B., Gstaettner, A., & Scherrer, P. (2021). Selfies to die for: A review of research on self-photography associated with injury/death in tourism and recreation. *Tourism Management Perspectives, 37*, 100778.

Wen, P. (2016, February 6). New year's revolution: Celebrating abroad. *Sydney Morning Herald*, 30.

Wen, H., Josiam, B. M., Spears, D. L., & Yang, Y. (2018). Influence of movies and television on Chinese Tourists perception toward international tourism destinations. *Tourism Management Perspectives, 28*, 211–219.

Westfall, S. (2017, October 13). Next stop… Yale: Chinese tourism on campus. *The Politic*. https://thepolitic.org/next-stop-yale-chinese-tourism-on-campus/

Wilk, V. (2019, June 21). How a humble Perth boathouse became Australia's most unlikely tourist attraction. *The Conversation*.

Wong, A., Anderson, K., Ang, I., & McNeill, D. (2016). *Sydney's Chinatown in the Asian Century: From ethnic enclave to global hub*. Institute for Culture and Society, Western Sydney University.

Wong, J., Lai, I., & Tao, Z. (2020). Sharing memorable tourism experiences on mobile social media and how it influences further travel decisions. *Current Issues in Tourism, 23*(14), 1773–1787.

Woodward, S. (2013). Campus tourism, universities and destination development. In I. M. Smith & G. Richards (Eds.), *The Routledge handbook of cultural tourism* (pp. 265–272). Routledge.

World Economic Forum. (2019). *The travel & tourism competitiveness report 2019: Travel and tourism at a tipping point*. World Economic Forum.

World Tourism Cities Federation & Ipsos. (2016). *Market research report on Chinese*.

Wright, T. (2019, April 27). 'Ghost' town booms as hordes hit remote lake to capture magic moments. *The Age*.

Wu, M. (2015). Driving an unfamiliar vehicle in an unfamiliar country: Exploring Chinese recreational vehicle tourists' safety concern and coping techniques in Australia. *Journal of Travel Research, 54*(6), 801–813.

Wu, M., & Pearce, P. (2014). Chinese recreational vehicle users in Australia: A netnographic study of tourist motivation. *Tourism Management, 43*, 22–35.

Wu, M., & Pearce, P. L. (2016). Tourism blogging motivations: Why do Chinese tourists create little 'Lonely Planets'? *Journal of Travel Research, 55*(4), 537–549.

Wu, B., Xue, L., Morrison, A. M., & Leung, X. Y. (2012). Frame analysis on golden week policy reform in China. *Annals of Tourism Research, 39*(2), 842–862.

Xia, J., Adriano, R., Zea, M., Crowe-Delaney, L., et al. (2018). *Are we China ready? Chinese tourism in Western Australia*. BCEC Research Report No. 14/18.

Xiang, Z., & Gretzel, U. (2010). Role of social media in online travel information search. *Tourism Management, 31*, 179–188.

Xie, L. (2016, August 23). How nice is the quad? *Honi Soit*.

Xu, H., & Wu, Y. (2016). Lifestyle mobility in China: Context, perspective and prospects. *Mobilities, 11*(4), 509–520.

Xu, H., Ding, P., & Packer, J. (2008). Tourism research in China: Understanding the unique cultural contexts and complexities. *Current Issues in Tourism, 11*(6), 473–491.

Yan, G. (2016) Chinese tourists remain 'explosive buying' but preferences changed, Thailand becomes the most popular. *China Youth News*, July 28, 2017. [In Chinese].

Yin, J., Brown, G., & Huang, S. (2020). Social media and Chinese tourist behaviour. In S. Huang & G. Chen (Eds.), *Handbook on tourism and China* (pp. 91–105). Elgar.

Yingqun, C. (2017, March 23). Tourists take off the shackles and begin to do their own thing. *China Watch*, 2.

Yoo, K., & Gretzel, U. (2011). Influence of personality on travel-related consumer-generated media creation. *Computers in Human Behavior, 27*(2), 609–621.

Yousaf, S., & Fan, X. (2020). Copysites/duplitectures as tourist attractions: An exploratory study on experiences of Chinese tourists at replicas of foreign architectural landmarks in China. *Tourism Management, 81*, 104179.

Yu, J., & Egger, R. (2021). Color and engagement in touristic Instagram pictures: A machine learning approach. *Annals of Tourism Research, 89*, 103204.

Yu, C., Xie, S., & Wen, J. (2020). Coloring the destination: The role of color psychology on Instagram. *Tourism Management, 80*, 104110.

Yu, H., Xu, J., & Sun, P. (2022). Introduction: Platformization of entrepreneurial labor via Chinese digital networks. *Global Media and China, 7*(3), 253–262.

Yuan, B., & Peluso, A. (2023, March 13). Travelling abroad or domestically? The effects of belief in karma and anxiety on post-pandemic travelling intention among Chinese tourists. *Current Issues in Tourism*.

Yuming, Y. (2018, July 6). Personal summer tour to Peking University Tsinghua needs online reservation. *Beijing News*. http://www.bjnews.com.cn/feature/2018/07/06/493977.html

Zarezadeh, Z., Rastegar, H., & Gretzel, U. (2018). Reviewing the past to inform the future: A literature review of social media in tourism. *Czech Journal of Tourism, 7*(2), 115–131.

Zeng, B., & Gerritsen, R. (2014). What do we know about social media in tourism? A review. *Tourism Management Perspectives, 10*, 27–36.

Zhang, L. (2017, August 23). 430,000 people visited Peking University, Tsinghua University during the summer, 30 of whom were detained for disturbing order. *The Legal Evening News*. https://www.thepaper.cn/newsDetail_forward_1772573

Zhang, X. (2022). Presentation, alienation and reconstruction of female media image in modern network live broadcast. *Advances in Journalism and Communication, 10*, 336–354.

Zhang, Y., & Murphy, P. (2009). Supply-chain considerations in marketing underdeveloped regional destinations: A case study of Chinese tourism to the Goldfields region of Victoria. *Tourism Management, 30*, 278–287.

Zhang, Y., & Peng, Y. (2014). Understanding travel motivations of Chinese tourists visiting Cairns. *Journal of Hospitality and Tourism Management, 21*, 44–53.

Zhang, X., & Ryan, C. (2018). Investigating tourists' and local residents' perceptions of a Chinese film site. *Tourist Studies, 18*(3), 275–291.

Zhang, J., & Shelton, E. (2015). Ordering the disordered subject: A critique of Chinese outbound tourists as New Zealand seeks to become China Ready. *Tourism Analysis, 20*, 343–353.

Zhang, C., Fong, L., & Li, S. (2019). Co-creation experience and place attachment: Festival evaluation. *International Journal of Hospitality Management, 81*, 193–204.

Zhang, Y., Zhang, C., & Yarde, K. (2021). Characteristics of China's air transport industry. In C. Zhang & K. Yarde (Eds.), *China's trade policy on international air transport: Policy goals, driving forces, and impact* (pp. 24–52). Routledge.

Zhang, H., Weng, S., & Bao, J. (2022). How do Chinese tourists negotiate the constraints of engaging in post-COVID-19 domestic travel? *Annals of Tourism Research Empirical Insights, 3*(2), 100065.

Zhang, H., Huang, K., Feng, W., & Zhu, J. (2023). Authentic well-being in the third space among Hanfu-wearing Chinese tourists: A PERMA analysis. *Tourism Management Perspectives, 47*, 1011106.

Zhong, L., Yang, L., Morrison, A. et al. (2023, April 14). Who wears Hanfu clothing when they travel? Experience-based cultural tourism market segmentation – The case of Xi'an, China. *Current Issues in Tourism.*

Zhou, F. (2009). Analysis of the current development of campus tourism and counter measures. *Contemporary Education Theory and Practice, 1*(5), 154–156.

Zhou, M., Lu, X., & Wang, B. (2023). Travel pictures authenticity in social media space. *Annals of Tourism Research, 101*, 103581.

Zhu, D., Xu, H., & Jiang, L. (2022). Behind buying: The Chinese gaze on European commodities. *Asia Pacific Journal of Tourism Research, 213*, 293–311.

Zhu, Z., Zhang, A., Zhang, Y., Huang, Z., & Xu, S. (2019). Measuring air connectivity between China and Australia. *Journal of Transport Geography,* 74, 359–370.

Zhuang, Y., & Everett, A. (2018). A brief history of Chinese wedding and bridal photography tourism: Through the lens of top Chinese wedding photographers. In E. Yang & C. Khoo-Lattimore (Eds.), *Asian cultures and contemporary tourism* (pp. 79–100). Springer Nature.

Zong, Y., & Tsaur, S. (2023, September 20). Destination experience scape for Hanfu Tourism: An explanatory study, *Journal of China Tourism Research.*

Zou, Y., & Yu, Q. (2022). Sense of safety towards tourism destinations: A social constructivist perspective. *Journal of Destination Marketing and Management, 24*, 1007708.

Index

J. Connell et al., *Chinese Tourism in Australia*,
https://doi.org/10.1007/978-981-97-2477-2

SPRINGER NATURE

GPSR Compliance

The European Union's (EU) General Product Safety Regulation (GPSR) is a set of rules that requires consumer products to be safe and our obligations to ensure this.

If you have any concerns about our products, you can contact us on ProductSafety@springernature.com

In case Publisher is established outside the EU, the EU authorized representative is:

Springer Nature Customer Service Center GmbH
Europaplatz 3
69115 Heidelberg, Germany

The manufacturer's authorised representative in the EU is Springer Nature Customer Service Centre GmbH, Europaplatz 3, 69115 Heidelberg, Germany. If you have any concerns regarding our products, please contact ProductSafety@springernature.com

Printed and bound by CPI Group (UK) Ltd, Croydon, CR0 4YY

30/09/2026

02225254-0002